Molecular Interactions and
Activity in Proteins

*The Ciba Foundation for the promotion of international cooperation in
medical and chemical research is a scientific and educational charity established by
CIBA Limited — now CIBA-GEIGY Limited — of Basle. The Foundation operates
independently in London under English trust law.*

*Ciba Foundation Symposia are published in collaboration with
Excerpta Medica in Amsterdam*

Excerpta Medica, P.O. Box 211, Amsterdam

Molecular Interactions and Activity in Proteins

Ciba Foundation Symposium 60 (new series)

1978

Excerpta Medica

Amsterdam · Oxford · New York

© *Copyright 1978 Ciba Foundation*

ISBN 0-444-90040-3
ISBN 90-219-4066-3
Published in November 1978 by Excerpta Medica, P.O. Box 211, Amsterdam and Elsevier/ North-Holland Inc., 52 Vanderbilt Avenue, New York, N.Y. 10017.

Suggested series entry for library catalogues: Ciba Foundation Symposia.
Suggested publisher's entry for library catalogues: Excerpta Medica.

Ciba Foundation Symposium 60 (new series)

287 pages, 99 figures, 23 tables

Library of Congress Cataloging in Publication Data

Symposium on Molecular Interactions and Activity in Proteins, London, 1977.
    Molecular interactions and activity in proteins.

    (Ciba Foundation Symposium; 60 (new ser.))
    Bibliography: p.
    Includes indexes.
    1. Proteins-Congresses. 2. Enzymes-Congresses. 3. Molecular association-Congresses. I. Title.
    II. Series: Ciba Foundation. Symposium; new ser., 60.
QP551.S936 1977   547.1'9245   78-14500
ISBN 0-444-90040-3

Printed in The Netherlands by Van Gorcum, Assen

Contents

Participants

Symposium on Molecular Interactions and Activity in Proteins, *held at the* Ciba Foundation, London, 6th and 7th December, 1977

A. R. BATTERSBY (*Chairman*)   University Chemical Laboratory, Lensfield Road, Cambridge CB2 1EW, UK

D. ARIGONI   Laboratorium für Organische Chemie, Eidgenössische Technische Hochschule, Universitätstrasse 16, CH-8006 Zürich, Switzerland

D. BLOW   Blackett Laboratory, Department of Physics, Imperial College of Science and Technology, Prince Consort Road, London SW7 2 BZ, UK

L. A. BLUMENFELD*   Institute of Chemical Physics, Academy of Sciences of the USSR, Vorob'evskoe Shosse 2-b, Moscow 117334, USSR

T. L. BLUNDELL   Department of Crystallography, Birkbeck College, Malet Street, London EC1E 7HX, UK

C.-I. BRÄNDÉN   Department of Chemistry, Division 1, The Swedish University of Agricultural Sciences, S-750 07 Uppsala, Sweden 7

SIR JOHN CORNFORTH and LADY CORNFORTH   School of Molecular Sciences, University of Sussex, Falmer, Brighton BN1 9QJ, UK

J. D. DUNITZ   Laboratorium für Organische Chemie, Eidgenössische Technische Hochschule, Universitätstrasse 16, CH-8006 Zürich, Switzerland

F. FRANKS   Department of Botany, University of Cambridge, Downing Street, Cambridge CB2 3EA, UK

H. GUTFREUND   Department of Biochemistry, University of Bristol, Bristol BS8 1TD, UK

* Unable to be present at the symposium.

M. J. HALSEY   Clinical Research Centre, Anaesthesia Division, Northwick Park Hospital, Watford Road, Harrow, Middlesex HA1 3UJ, UK

J. T. JOHANSEN   Carlsberg Laboratory, Chemical Department, Gamle Carlsberg Vej 10, DK-2500 Copenhagen Valby, Denmark

G. W. KENNER*   The Robert Robinson Laboratories, University of Liverpool, Oxford Street, PO Box 147, Liverpool L69 3BX, UK

J. R. KNOWLES   Department of Chemistry, Harvard University, Cambridge, Massachusetts 02138, USA

W. N. LIPSCOMB   Gibbs Chemical Laboratory, Harvard University, Cambridge, Massachusetts 02138, USA

D. C. PHILLIPS   Laboratory of Molecular Biophysics, Department of Zoology, South Parks Road, Oxford OX1 3PS, UK

F. M. RICHARDS   Department of Molecular Biophysics and Biochemistry, Yale University, New Haven, Connecticut 06520, USA

G. C. K. ROBERTS   National Institute for Medical Research, Mill Hill, London NW7 1AA, UK

B.-M. SJÖBERG   Medical Nobel Institute, Department of Biochemistry I, Karolinska Institutet, Solnavägen 1, S-104 01 Stockholm, Sweden

R. M. TOPPING   School of Molecular Sciences, University of Sussex, Falmer, Brighton BN1 9QJ, UK

B. L. VALLEE   Biophysics Research Laboratory, Department of Biological Chemistry, Harvard Medical School, Boston, Massachusetts 02115, USA

R. J. P. WILLIAMS   Department of Inorganic Chemistry, Oxford University, Oxford OX1 3QR, UK

K. WÜTHRICH   Institut für Molekularbiologie und Biophysik, Eidgenössische Technische Hochschule Zürich-Hönggerberg, CH-8049 Zürich, Switzerland

Editors: RUTH PORTER (*Organizer*) and DAVID W. FITZSIMONS

* Died, July 1978.

Intramolecular interactions, enzyme activity and models

WILLIAM N. LIPSCOMB

Gibbs Chemical Laboratory, Cambridge, Massachusetts

Abstract The specificity of protein binding and the specificity–catalysis relationship in enzymes are analysed. Many enzymes use extended binding sites to achieve specificity and to create special environments which activate chemical groups on both the substrate and the enzyme itself. Aside from this effect, which is not adequately available in model compounds, the use of models is exemplified for several possibly separable effects in enzyme–substrate reactions. These include proximity (entropy loss on binding), locking into a productive binding mode, desolvation, electrostatic effects, changes of pK_a by local environments, geometric strain, acid-base catalysis, and formation of other intermediates.

Proteins interact with other proteins, with small molecules and with ions. The transient or stable complexes thus formed often have a biochemical function. Examples of function are transport, stabilization, conversion of the protein or substrate into another form, or chemical transformation of the species bound or of active groups of the enzyme. Many proteins have a definite, if flexible, tertiary three-dimensional structure which conveys specificity, special environments, enhanced reactivities of substrates or groups on the protein, and other properties.

Enzymes form a special class of proteins which catalyse certain reactions. Even though they participate in the chemical steps of the reaction, they finally emerge unchanged, ready to begin again their action upon a new substrate molecule which is converted into products. These reactions are usually specific. Their rates exceed those of the uncatalysed reaction by factors of 10^8 to 10^{14} or higher for 1 mol/l standard states, and by even greater factors if more realistic states of 10^{-6} mol/l are used. Most non-enzyme-catalysed reactions are some 10^2–10^6 times faster than the uncatalysed reactions, although some model reactions discussed below show considerably larger rate enhancements.

1

Why are enzymes so specific and why do they catalyse a reaction so efficiently? If there were a single cause, it would have long since been described adequately. Of the several factors discussed below we shall see that the enzyme–substrate interactions and transformations occur by mechanisms already known to the organic chemist. In addition, the specific binding interactions are used to overcome the less favourable aspects of corresponding uncatalysed reactions (Westheimer 1962; Jencks 1975).

SPECIFICITY AND RATE ENHANCEMENTS

It is the intimate relationship between activity and specificity that characterizes the great efficiency of enzymic catalysis. The specific binding interactions at an allosteric site, at the active site, and near the active site bring the catalytic groups of the enzyme (or substrate) into precise position, create an environment which activates those groups and produce various strains on the ES complex which is then transformed along the reaction pathway. Allosteric sites are omitted here, as are the ubiquitous steric interactions of a substrate which cannot be fitted into an active site of the enzyme.

There are two aspects of the 'induced fit' proposal (Koshland 1958; Koshland & Neet 1968; Citri 1967). The first is exemplified by geometric strain, for which the clearest early statement was given by Pauling (1948): 'I think that enzymes are molecules that are complementary in structure to the activated complexes of the reaction that they catalyse.' Of course, both enzyme and substrate are transformed in their mutual interaction so that the transition state toward products is approached (Fig. 1). Carboxypeptidase A and lysozyme illustrate this aspect of enzyme–substrate interaction well. Estimates of the rate enhancements due to this factor are shown in Fig. 2 for two inhibitors (tris[N-acetylglucosamine]-xylosamine and -lactone) which are not highly strained at the fourth ring of the polysaccharide-like substrate (Ford et al. 1974). The estimate of -5.2 kcal/mol for binding of the gluconolactone at subsite D may be a bit high. Secemski & Lienhard (1971) estimate from a study of this inhibitor that relief of strain contributes a factor of 10^3–10^4 in catalysis by lysozyme. However, see Schindler et al. (1977).

The second aspect of induced fit arises from the extended nature of the binding site. For example, hexokinase phosphorylates glucose about 4×10^4 times faster than it phosphorylates water (Jencks 1975). The interactions of the glucose molecule with hexokinase are more extensive than those of water and cause a change in the enzyme from an inactive conformation (E) to an active form (E'), as summarized in Fig. 3. Part of the binding (free) energy of glucose is thus utilized in shifting the E $\leftrightharpoons$ E' equilibrium so that the active

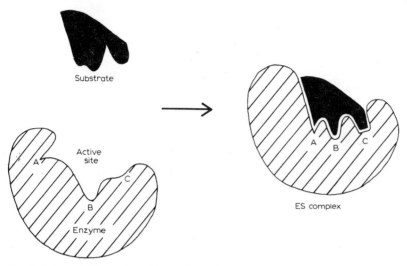

FIG. 1. Schematic illustration of the conformational transformation of both enzyme and substrate upon interaction to form an approximation to the transition state (ES) in the pathway from substrate to product.

form is in greater relative concentration. These interactions occur in the region of the active site and are usually distinct from geometric strain at the critical bonds in the reaction itself. These examples of specificity are clearly inseparable from the rate process of the enzyme-substrate reaction.

One result is that the observed binding energy, typically -5 kcal/mol, of the ES complex is considerably less than that seen in, for example, an antigen–antibody interaction ($K_s = 10^{13}$ l/mol, equivalent to a binding energy of some -18 kcal/mol). If one estimates $+8$ kcal/mol for losses of translational

ΔG, kcal/mol at subsites in lysozyme					
A	B	C	D		
-1.8	-3.7	-5.7	$+2.9$	(NAG)n = 1, 2, 3, 4	Rate enhancement
			-2.2	(NAG)$_3$ xylosamine	$\sim 10^4$
			-5.2	(NAG)$_3$ lactone	$\sim 10^6$
			-5.7	If like subsite C	$\sim 10^6$

FIG. 2. Estimates of rate enhancements due to strain at the fourth subsite (ring D) of lysozyme when tetrasaccharides or analogues are bound in nearly the normal mode. The geometric strain which occurs at subsite D when N-acetylglucosamine (NAG) is bound is relieved when the xylosamine or gluconolactone analogue of the transition state is substituted, or when one guesses that the binding due to other factors than strain is the same at subsites C and D. For other ground- and transition-state effects see Schindler et al. (1977).

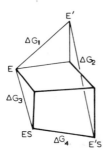

FIG. 3. Substrate-induced activation of hexokinase (E) by glucose (S).

$$E \underset{}{\overset{K_1}{\rightleftharpoons}} E' \text{ (active)}$$

$$K_3 \Big\updownarrow \pm S \qquad \pm S \Big\updownarrow K_2$$

$$ES \underset{K_4}{\rightleftharpoons} E'S$$

$$K_1 K_2 = K_3 K_4$$

ΔG_3 from $K_s = 2 \times 10^{-4}$ mol/l for glucose dissociation. ΔG_1 from $K_1 = 2.5. \times 10^{-5}$ so H_2O not phosphorylated. Thus ΔG_2 is more than twice the observed binding free energy: $RT\ln(2 \times 10^{-4}) + RT\ln(2.5 \times 10^{-5}) = -11$ kcal/mol.

Entropy changes, ΔS				
(a) Gas	A +	B ⇌	AB	ΔS/e.u.
S trans	34	34	35	−33
S rot	25	25	28	−22
S internal	3	3	18	12
S total	62	62	81	−43
−ΔS vap	↓−10	↓−10	↓−15	
(b) Solution	52	52	66	−38
(c) Degrees of freedom				
Translation	3	3	3	∼ 10^8 mol/l
Rotation	3	3	3	
Vibration	$3n-6$	$3n'-6$	$3(n + n')-6$	

FIG. 4. Entropy changes in a bimolecular reaction and their relationship to losses of degrees of translational and rotational freedom in a bimolecular reaction (Jencks 1975).

and rotational freedom when the ES complex is formed, and +7 kcal/mol for destabilization due to geometric strain, electrostatic and dehydration effects, the 'intrinsic' binding energy is −5 −7 −8 = −20 kcal/mol. This is the free energy which would apply if the formation of the productive ES complex did not have to overcome the unfavourable entropic and energetic aspects of the reaction. The losses of degrees of freedom, summarized in Fig. 4, are shown as entropy changes. The value of 38 e.u. corresponds to a free-energy change of 11.4 kcal/mol at 300 K and to a rate enhancement of 10^8 mol/l for 1 mol/l standard state, assuming that the formation of the ES complex is not rate-limiting.

Factors by which rates are enhanced by various processes (entropy loss, strain, electrostatic modifications, desolvation, induced fit, productive binding) have been classified differently by the more vocal scientists in this field. A

	Approximation	Orientation factor
Koshland	≤55 mol/l	~ 10^4 centre[b]
Bruice	10^3 mol/l	230/rotation
Jencks	10^8 mol/l	10/rotation

[a] Concentration effect only
[b] Entropy losses
[c] Solvent reorganization elsewhere

FIG. 5. An attempt to classify the factors of rate enhancement from various effects in enzymic reaction from the studies of Koshland, Bruice and Jencks (see, e.g., Bruice 1970; Page & Jencks 1971).

summary (Fig. 5) may not adequately reflect the chronological changes of interpretation but it does indicate that all are basically right.

SEPARABLE EFFECTS?

The isomerases provide a simpler set of problems in which energetic and entropic effects are often associated with the isomerization steps only. Conversion of chorismate into prephenate (Fig. 6) is speeded up by a factor of 2×10^6 by the enzyme chorismate mutase-prephenate dehydrogenase (Andrews et al. 1973). Here, the observed ΔH^* (activation) is 21 kcal/mol and ΔS^* is −13 e.u. for the uncatalysed reaction. This enzymic rate enhancement (2×10^6) corresponds to a ΔG^* value of 9 kcal/mol. If ΔS^* is the same for the catalysed and the uncatalysed reactions, ΔH^* is reduced to 12 kcal/mol. Alternatively, if ΔS^* is zero for the catalysed reaction, ΔH^* is reduced by 5 kcal/mol to a value of 16 kcal/mol. Thus the enzyme reduces ΔH^* by some 5–9 kcal/mol. A recent study of transition-state analogues suggests that the chair-like intermediate with equatorial oxygen is favoured (Andrews et al. 1977). The relationship of this isomerase and the well studied triosephosphate isomerase (Knowles & Albery 1977) to the discussion below is that these systems are particularly suitable for force-constant and molecular-orbital studies of models of the reaction pathway.

The role of water in the function of proteins, especially enzymes, is not well understood. There is a comment below for model systems, but in this section on real enzymes perhaps one can say that much more can be learned about protein–substrate reactions and about protein denaturation and renaturation from more detailed studies of pressure effects. The volume changes associated with these processes are essentially different from those seen for interactions

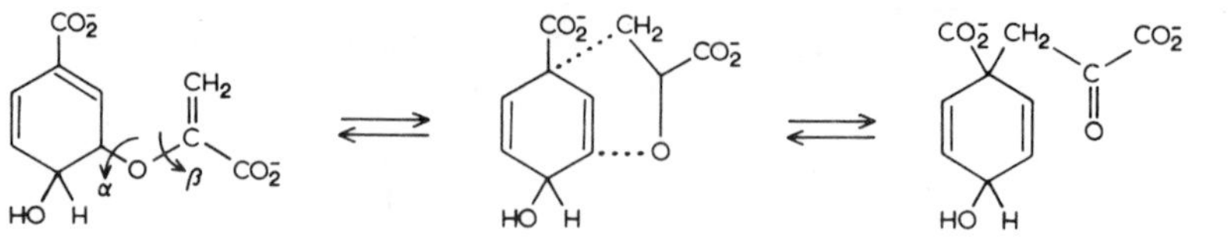

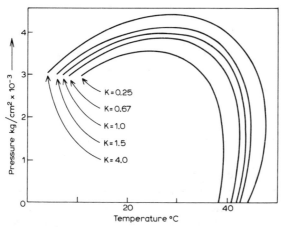

FIG. 6. The conversions of chorismate into prephenate by the enzyme chorismate mutase–pre-phenate dehydrogenase (EC 1.3.1.12) or chorismate mutase–prephenate dehydratase (EC 4.2.1.51). Uncatalysed: $\Delta H^* = 21$ kcal/mol, $\Delta S^* = -13$ e.u.

FIG. 7. Pressure–temperature diagram for the equilibrium constant K for the [denatured]/[native] ratio of chymotrypsinogen. Contours are shown for constant values of K (from Hawley 1971).

of small hydrophobic molecules with water. One example is the pressure dependence of the equilibrium between natured and denatured chymotryp-sinogen (Hawley 1971), summarized in Fig. 7. There are other examples and other methods for a more concerted attack on the problems of water in protein interactions. In our map of carboxypeptidase A, S. J. Wieland has located 686 water molecules of the expected 980 per half of the unit cell. Further refine-ment and filling of vacancies may yield a nearly complete description of this medium (0.1M-LiCl) in the crystals. Also residence times can be studied on the nuclear magnetic resonance time scale. Even so, this is a very difficult problem from either the experimental or theoretical point of view.

The effect of local environment in changing the pK_a of specific amino acid side-chains in a protein is at least more easily described (Fig. 8). One result is a great change in the concentration of an effective group at certain pH values.

ΔpK_a's	(pK_a)	ΔpK_a
Lys, acetoacetate decarboxylase	(6.0)	4.7
Cys-25, papain	(4.2)	4.3
Asp-102, chymotrypsin	(6.7)?	3.2?
Glu-270, carboxypeptidase A	(7)	2.5
His-159, papain	(8.6)	2.3
Glu-35, lysozyme	(6.6)	2.1
Cys-247, rhodanese	(6.5)	2.0
H_2O ZnL_3, carbonate dehydratase	(7)	2.0
Phenol of pyridoxal phospate as		
Schiff's base, aspartate aminotransferase	(6.2)	4.3

ΔpK_a of $1 \approx 1.4$ kcal/mol

FIG. 8. Changes of pK_a values due to environmental effects in enzyme structures.

For example, in acetoacetate decarboxylase the change of pK_a of the active site lysine from 10.7 to 6.0 by the presence of a nearby positive charge effectively increases the concentration of the neutral form of lysine by a factor of 10^4–10^5. This change increases the rate of formation of the Schiff's base with the substrate by about this factor (Frey *et al.* 1971; Kokesh & Westheimer 1971). A similar analysis can be made of many of the other changes of pK_a shown in Fig. 8. Although shifts of five units are highly unusual, shifts of pK_a by about two units are frequently observed.

The identification of oxidation states in ferredoxin and in high-potential iron protein (see Fig. 9; Carter *et al.* 1972) has been confirmed by the preparation of the superoxidized cluster in ferredoxin and the superreduced cluster in high-potential iron protein. The ability of the ferredoxin cluster to accommodate normally one more electron than the cluster in high-potential iron protein has been correlated with the considerably larger number of NH $\cdots$ S hydrogen bonds in the ferredoxin cluster (Adman *et al.* 1975).

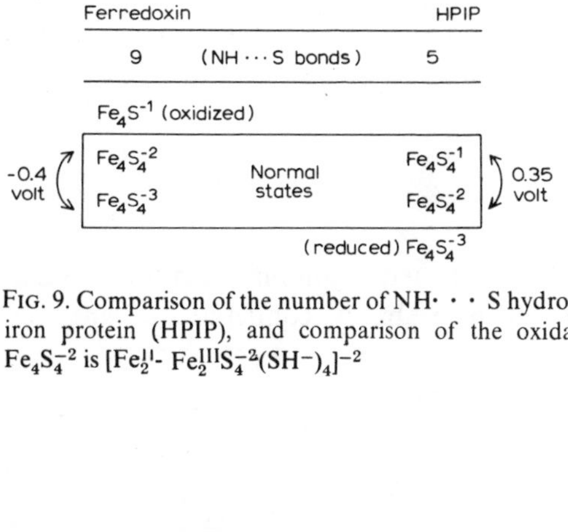

FIG. 9. Comparison of the number of NH $\cdots$ S hydrogen bonds in ferredoxin and high-potential iron protein (HPIP), and comparison of the oxidation–reduction states. $Fe_4S_4^{-2}$ is $[Fe_2^{II}\text{-}Fe_2^{III}S_4^{-2}(SH^-)_4]^{-2}$

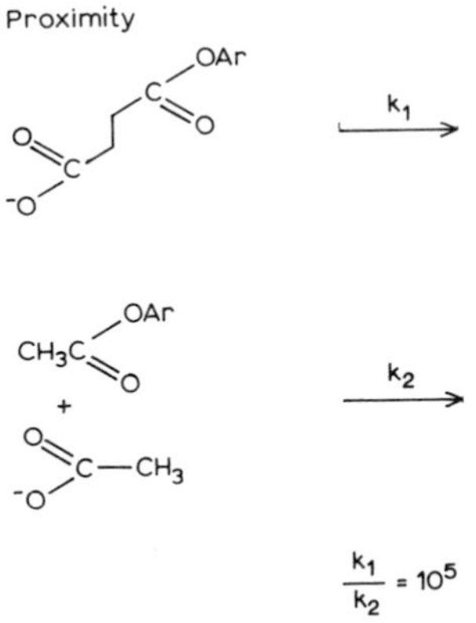

Proximity

$$\frac{k_1}{k_2} = 10^5$$

FIG. 10. Effect of proximity on a model reaction, in which there is still some degree of rotational freedom about the single bond in the reactant of the first example (Bruice & Pandit 1960).

Presumably, there is thus some preferential transfer of negative charge away from the cluster in ferredoxin.

The local environmental factors governing the redox potentials of the cytochromes are considerably more complex. Some of the factors are electrostatic charge on the ligand donor and acceptor power of the ligand, changes of spin state of the metal ion, and steric factors (Moore & Williams 1977). However, studies in my research group have yet to yield a relatively simple correlation of these and other effects with redox potentials.

A consistency of structure design for metal ligands has been noticed by D. Rees of my laboratory. Functional single metal ions are bound by widely separated protein ligands, the most distant pair separated by some 50 ±30 animo acid residues. Moreover, these ligands are usually on two or more regions of α-helix or β-strand. In contrast, structural metal ligands are almost always on a single loop of irregular secondary structure, which may also include NH_2- or COOH-terminal regions of adjacent structure. Overall separation of these ligands averages only 17 ± 13 residues. It therefore seems unlikely to Rees that the structural Zn site of liver alcohol dehydrogenase was a catalytic site at an earlier evolutionary stage, a possibility which was suggested earlier (Eklund et al. 1976).

MODELS

The following zoo of models has been selected as among the more striking examples of the separate factors which probably contribute to enzyme activity.

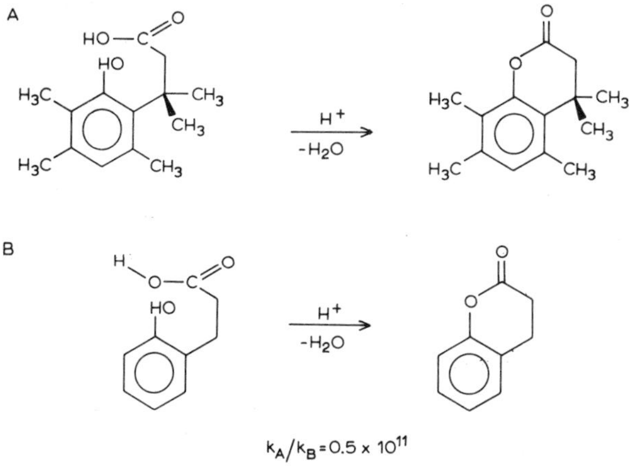

FIG. 11. Lock into position: the 'dimethyl lock' of Milstien & Cohen illustrates an extreme example of rate enhancement.

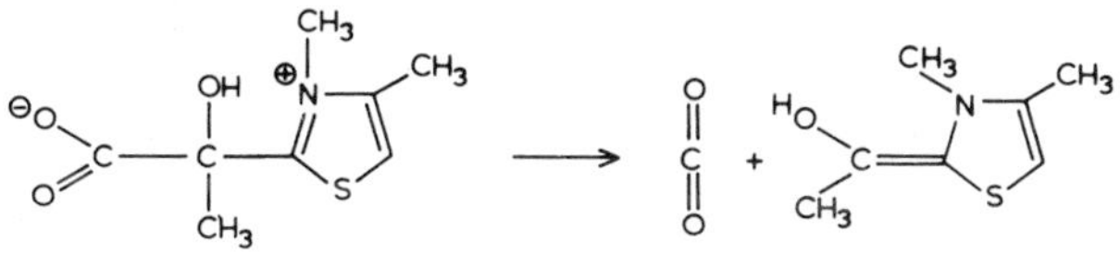

FIG. 12. Dehydration (from water to oil): decarboxylation of 2-(1-carboxy-1-hydroxyethyl)-3,4-dimethylthiazolium cation in the reactive zwitterion form (Crosby *et al.* 1970). 10^4–10^5 faster in ethanol than in H_2O. General: 10^4–10^8 increase in rate.

Proximity, in the covalent attachment of two reacting groups is worth a factor of 10^5 in the example shown in Fig. 10 (Bruice & Pandit 1960). Here, some degree of rotational freedom remains, but the effects of translational and some rotational freedom have been overcome by the covalent attachment of the reacting groups.

The locking of a substrate into final position so that it is pushed along the reaction pathway is nicely illustrated (Fig. 11) by the 'dimethyl lock' (Milstien & Cohen 1969, 1970, 1972). The rate enhancement of almost 10^{11} over that of the model without methyl groups is a striking result. Less certain is the comparison of this example where there are a few highly strained contacts with the enzyme–substrate complex where there are many more less-energetic contacts made up of dispersion, dipole, hydrogen-bond and salt-linked interactions.

Desolvation is illustrated by a model (Fig. 12), which decarboxylates some 10^4 – 10^5 times faster in ethanol than in water (Crosby *et al.* 1970). Presumably the charge transfer as the reactive zwitterion decarboxylates is promoted by a

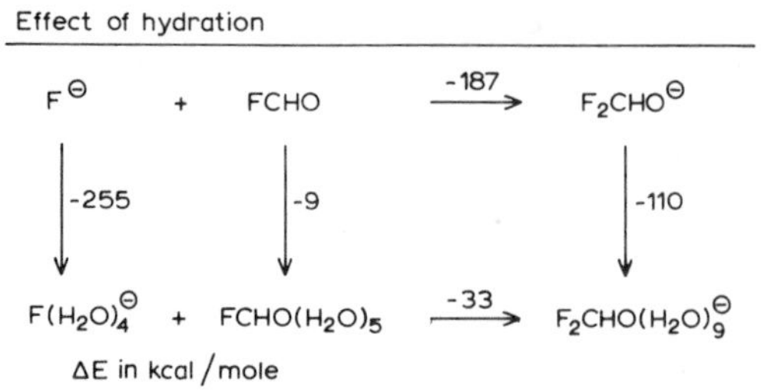

Effect of hydration

$$F^{\ominus} \quad + \quad FCHO \quad \xrightarrow{-187} \quad F_2CHO^{\ominus}$$

$$\downarrow -255 \qquad\qquad \downarrow -9 \qquad\qquad\qquad \downarrow -110$$

$$F(H_2O)_4^{\ominus} \quad + \quad FCHO(H_2O)_5 \quad \xrightarrow{-33} \quad F_2CHO(H_2O)_9^{\ominus}$$

ΔE in kcal/mole

FIG. 13. Theoretical study of the effect of a few water molecules on the formation of F₂CHO⁻ from F⁻ and FCHO (Scheiner *et al.* 1976).

non-aqueous environment. This is a model for the reaction promoted by pyruvate decarboxylase. Adequate theoretical methods for including solvent water into a model reaction have not been developed. An example of a large effect of a few water molecules is shown in Fig. 13 for a model reaction of nucleophilic attack (Scheiner *et al.* 1976). However, the number of water molecules is probably too small, unless they can be joined into a larger approximate structure by approximate methods (Warshel & Levitt 1976).

In lysozyme Asp-52 may stabilize a carbonium ion intermediate rather than make a covalent bond with the fragment of the substrate. A model comparison (Fig. 14) of the *ortho*-carboxylate- and *para*-carboxylate-phthal-aldehydic acid acetals shows that this stabilization is worth at least a factor of 100 (Fife & Przystas 1977). This effect may be enhanced by a hydrophobic

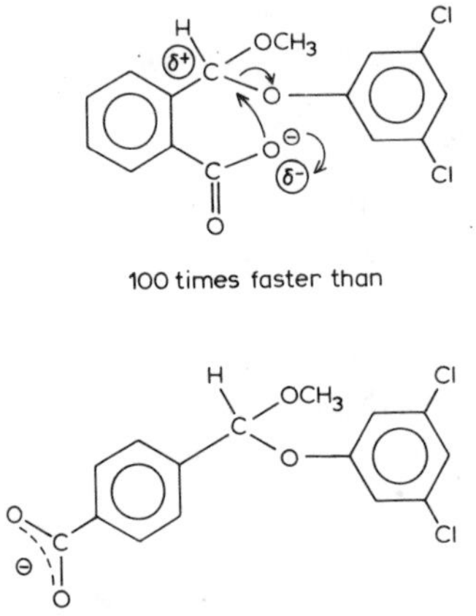

100 times faster than

FIG. 14. Stabilization of a developing carbonium ion by a carboxylate anion in phthalaldehydic acid methyl 3, 5-dichlorophenyl acetal (Fife & Przystas 1977).

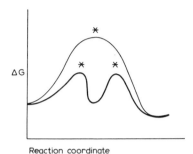

FIG. 15. Schematic illustration of barrier reduction by the formation of an intermediate in a reaction. Barrier reduction 5.5 kcal/mol; rate enhancement 10^4.

environment. Hence, this model reaction should be studied in a solvent of low dielectric constant.

Enzymes do chemistry. They often form defined intermediates, which convert a high barrier into a succession of lower barriers (Fig. 15). General acid-base catalysis and the formation of a covalent intermediate such as an acylenzyme illustrate this mechanism of rate enhancement. A striking example is general acid catalysis of acetal hydrolysis by a carboxylate group (Anderson & Fife 1973) as shown in Fig. 16. General acid catalysis occurs in several hydrolytic enzymes, for example in lysozyme where Glu-35 performs this function.

In a model of the chymotrypsin mechanism, in which Ser-195 is taken as methanol, His-57 as imidazole, Asp-102 as formate anion and the substrate as formamide, concerted proton transfers from Ser-195 to His-57 and from His-57 to Asp-102 were found energetically preferable to consecutive proton

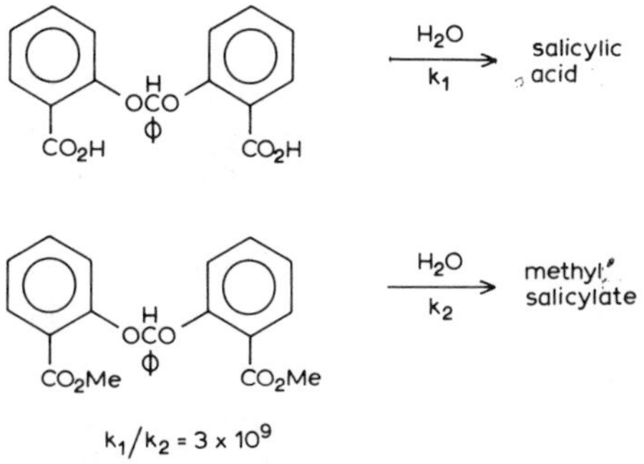

FIG. 16. General acid catalysis of a disalicyl acetal by a carboxylate group enhances the rate by a factor of more than 10^9 (Anderson & Fife 1973).

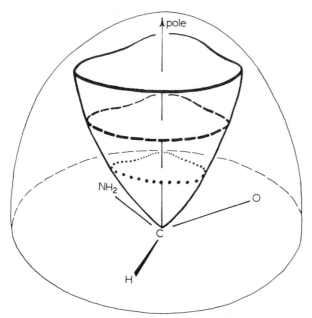

Fɪɢ. 17. The reaction cone for zero activation energy when methoxide anion attacks the carbon atom of formamide. Within the polar angle, measured from the axis, of 37° (20% of the solid angle) the activation energy is zero or less. Approximately the same numerical results were obtained when the oxygen atom of methanol attacks the carbon atom of formamide.

transfers. Also, the Ser O⁻ ion attacks after some 80% of the concerted proton transfers take place. General acid-base catalysis was found to lower the barrier for hydrolysis by about 16 kcal/mol. However, the resulting barrier of some 33 kcal/mol is still too high by about a factor of two or three (Scheiner *et al.* 1975; Scheiner & Lipscomb 1976). This kind of molecular-orbital model, even though it starts from the X-ray coordinates and passes through the coordinates approaching the transition state (Huber *et al.* 1974), is inadequate because of neglect of other stabilizing influences on the enzyme, neglect of solvent, and restriction so far to a minimum basis set of atomic orbitals. However, it is a promising start and will be improved upon. One must be careful not to ask for too much, perhaps only for the sequential order of the steps, and not (at least, not yet) for a good account of the energetics. A separate study of the angle of approach of the model nucleophile CH_3O^- to the carbon atom of formamide yields the reaction cone shown in Fig. 17 (Scheiner *et al.* 1976). One should recognize that this angle of approach is only one of the factors involved in the concept of 'orbital steering'.

In summary, the value of these models is to show that there are more than enough rate-enhancement mechanisms available to account for the remarkably high rates of enzyme-catalysed reactions. However, it is a difficult task for the future to analyse these factors, if indeed they are separable, as they apply to any given enzyme–substrate system. Even so, model systems do not adequately account for the use of extended binding to shift from inactive to active conformations, nor are solvent effects and local hydrophobic or changed environments well treated in the models that are presently available. No one has yet synthesized a molecular catalyst that shows both the specificity and the rate-enhancement factors of a moderately effective enzyme!

References

ADMAN, E., WATENPAUGH, K. D. & JENSEN, L. (1975) NH···S hydrogen bonds in *Streptococcus aerogenes* ferredoxin, *Clostridium pasteurianum* rubridoxin, and *Chromatium* high potential iron protein. *Proc. Natl. Acad. Sci. U.S.A. 72*, 4854-4858

ANDERSON, E. & FIFE, T. H. (1973) Carboxyl group participation in acetyl hydrolysis. Hydrolysis of disalicyl acetals. *J. Am. Chem. Soc. 95*, 6437-6441

ANDREWS, P. R., SMITH, G. D. & YOUNG, I. E. (1973) Transition state stabilization and enzymic catalysis. Kinetic and molecular orbital studies of the rearrangement of chorismate to prephenate. *Biochemistry 12*, 3492-3498

ANDREWS, P. R., CAIN, E. N., RIZZARDO, E. & SMITH, G. D. (1977) Rearrangement of chorismate to prephenate. Use of chorismate mutase inhibitions to define the transition state structure. *Biochemistry 16*, 4848-4852

BRUICE, T. C. (1970) Proximity effects and enzyme catalysis, in *The Enzymes,* 3rd edn., vol. II, ch. 4 (P. D. Boyer, ed.), p. 217, Academic Press, New York

BRUICE, T. C. & PANDIT, U. K. (1960) The effect of geminal substitution, ring size and rotamer distribution on the intramolecular nucleophilic catalysis of the hydrolysis of monophenyl esters of dibasic acids and the solvolysis of the intermediate anhydrides. *J. Am. Chem. Soc. 82*, 5858-5865

CARTER, C. W., KRAUT, J., FREER, S. T., ALDEN, R. A., SIEKER, L. C., ADMAN, E. & JENSEN. L. H. (1972) A comparison of Fe_4S_4 clusters in high potential iron protein and in ferredoxin. *Proc. Natl. Acad. Sci. U.S.A. 69*, 3526-3529

CITRI, N. (1967) Conformational adaptability in enzymes. *Adv. Enzymol. 37*, 397-648

CROSBY, J., STONE, R. & LIENHARD, G. E. (1970) Mechanisms of thiamine-catalyzed reactions. Decarboxylation of 2-(1-carboxy-1-hydroxyethyl)-3,4-dimethylthiazolium chloride. *J. Am. Chem. Soc. 92*, 2891-2900

EKLUND, H., NORDSTROM, B., ZEPPEZAUER, E., SÖDERBERG, G., OHLSSON, I., BOIWE, T., SODERBERG, B.-O., TAPDIR, O. & BRÄNDEN, C.-I. (1976) Three-dimensional structure of horse liver alcohol dehydrogenase at 2.4 Å resolution. *J. Mol. Biol. 102*, 27-59

FIFE, T. H. & PRZYSTAS, T. J. (1977) Carboxylate anion stabilization of a developing carbonium ion in acetal hydrolysis. Hydrolysis of phthalaldehyde acid acetals. *J. Am. Chem. Soc. 99*, 6693-6699

FORD, L. O., JOHNSON, L. N., MACHIN, P. A., PHILLIPS, D. C. & TJIAN, R. (1974) Crystal structure of a lysozyme–tetrasaccharide lactone complex. *J. Mol. Biol. 88*, 349-371

FREY, P. A., KOKESH, F. C. & WESTHEIMER, F. H. (1971) A reporter group at the active site of acetoacetate decarboxylase. I. Ionization constant of the nitrophenol. *J. Am. Chem. Soc. 93*, 7266-7269

HAWLEY, S. A. (1971) Reversible pressure–temperature denaturation of chymotrypsinogen. *Biochemistry 10*, 2436-2442

HUBER, R., KUKLA, D., BODE, W., SCHWAGER, P., BARTELS, K., DIESENHOFER, J. & STEIGEMANN, W. (1974) Structure of the complex formed by bovine trypsin and bovine pancreatic trypsin inhibitor. II. Crystallographic refinement at 1.9 Å resolution. *J. Mol. Biol. 89*, 73-101

JENCKS, W. P. (1975) Binding energy, specificity, and enzymic catalysis: the Circe effect. *Adv. Enzymol. 43*, 219-410

KNOWLES, J. R. & ALBERY, W. J. (1977) Perfection in enzyme catalysis: the energetics of triosephosphate isomerase. *Acc. Chem. Res. 10*, 105-111

KOKESH, F. C. & WESTHEIMER, F. H. (1971) A reporter group at the active site of acetoacetate decarboxylase. I. Ionization constant of the amino group. *J. Am. Chem. Soc. 93*, 7270-7274

KOSHLAND, D. E. JR. (1958) Applications of a theory of enzyme specificity to protein synthesis. *Proc. Natl. Acad. Sci. U.S.A. 44*, 98-104

KOSHLAND, D. E. JR. & NEET, K. E. (1968) The catalytic and regulatory properties of enzymes. *Annu. Rev. Biochem. 37*, 359-410

MILSTIEN, S. & COHEN, L. A. (1969) Concurrent general-acid and general-base catalysis of esterification. *J. Am. Chem. Soc. 91*, 4585-4587

MILSTIEN, S. & COHEN, L. A. (1970) Concurrent general-acid and general-base catalysis of esterification. *J. Am. Chem. Soc. 92*, 4377-4382

MILSTIEN, S. & COHEN, L. A. (1972) Stereopopulation control. I. Rate enhancement in the lactonizations of *o*-hydroxyhydrocinnamic acids. *J. Am. Chem. Soc. 94*, 9158-9165

MOORE, G. R. & WILLIAMS, R. J. P. (1977) Structural basis for the variation in redox potential of cytochromes. *FEBS (Fed. Eur. Biochem. Soc.) Lett. 79*, 229-232

PAGE, M. I. & JENCKS, W. (1971) Entropic contributions to rate acceleration in enzymic and intramolecular reactions and the chelate effect. *Proc. Natl. Acad. Sci. U.S.A. 68*, 1678-1683

PAULING, L. (1948) Nature of forces between large molecules of biological interest. *Nature (Lond.) 161*, 707-709

SCHEINER, S. & LIPSCOMB, W. N. (1976) Molecular orbital studies of enzyme activity: catalytic mechanism of serine proteinases. *Proc. Natl. Acad. Sci. U.S.A. 73*, 432-436

SCHEINER, S., KLEIER, D. A. & LIPSCOMB, W. N. (1975) Molecular orbital studies of enzyme activity. The charge relay system and tetrahedral intermediate in the acylation of serine proteinases. *Proc. Natl. Acad. Sci. U.S.A. 72*, 2606-2610

SCHEINER, S., LIPSCOMB, W. N. & KLEIER, D. (1976) Molecular orbital studies of enzyme activity. 2. Nucleophilic attack on carbonyl systems with comments on orbital steering. *J. Am. Chem. Soc. 98*, 4770-4777

SCHINDLER, M., ASSAF, Y., SHARON, N. & CHIPMAN, D. M. (1977) Mechanism of lysozyme catalysis: role of ground-state strain in subsite D in hen egg-white and human lysozymes. *Biochemistry 16*, 423-431

SECEMSKI, I. I. & LIENHARD, G. E. (1971) The role of strain in catalysis by lysozyme. *J. Am. Chem. Soc. 93*, 3549-3550

WARSHEL, A. & LEVITT, M. (1976) Theoretical studies of enzymic reactions. Dielectric, electrostatic and steric stabilization of the carbonium ion in the reaction of lysozyme. *J. Mol. Biol. 103*, 227-249

WESTHEIMER, F. H. (1962) Mechanisms related to enzyme catalysis. *Adv. Enzymol. 24*, 441-482

Discussion

Blow: What effects might water have on your calculations about the active site of chymotrypsin?

Lipscomb: In a model in which one water molecule was hydrogen-bonded to the carbonyl oxygen atom of the scissile peptide bond, the stabilization was 25 kcal/mol greater for the tetrahedral intermediate than for the substrate (Scheiner & Lipscomb 1976). However, the conclusion that about 80% of the coordinated proton transfer takes place before substantial electrophilic attack occurs is unchanged.

Blow: Were you able to change the bulk dielectric constant around the active site?

Lipscomb: We did not account for the bulk dielectric constant. There are, however, more general methods, such as that of Warshel & Levitt (1976). Also M. D. Newton has developed a self-consistent field theory of a discrete molecular aggregate embedded in a continuum. These methods, still somewhat undeveloped, may be useful here.

The particular hydrogen-bond referred to in my first comment simulates the two hydrogen-bonded interactions from the enzyme to the carbonyl oxygen atom of the substrate (Rühlmann *et al.* 1973). Interactions with many more water molecules should be studied, but we are not able to afford energy-minimization calculations for these larger models.

Franks: There are dangers in extrapolating from hypothetical *in vacuo* states of hydrated proteins to the dilute solution. Quite apart from the weaknesses inherent in the calculation of hydration and conformation of an isolated protein, there is a fundamental principle at stake. Experimental studies are normally made at constant temperature and pressure, and weight concentrations are used (e.g. molality); i.e. we are dealing with the (T, P, m) ensemble. This gives us the Gibbs free energy, enthalpy, heat capacity at constant pressure, etc.

Statistical mechanical and conformational calculations are done with the condition of constant volume, i.e. the (T, V, c) ensemble where c is the concentration in weight per unit volume. This gives us the Helmholtz free energy, internal energy, heat capacity at constant volume, etc. It is this *latter* approach which gives direct information about interactions between molecules, but this is not generally recognized. One frequently comes across references to 'binding', based on ΔG or ΔH measurements when ΔA or ΔU should have been measured. One can readily make the transformation from the (T, P, m) to the (T, V, c) ensemble, provided the necessary volumetric data are available. The PdV term (included in ΔG and ΔH) can be large, as Professor Lipscomb showed for chymotrypsin (Fig. 7). There are even cases where ΔG and ΔA have opposite signs, and what is then believed to be an attraction (based on ΔG) is in fact a repulsion.

Lipscomb: I agree. I mentioned both free energies, stressing the pressure dependence and raising questions about the energy of the reaction. In a model reaction one can study an entropy effect by varying the temperature. Enzyme reactions involve too many other effects for the use of this simple procedure.

Franks: But in model experiments with little extra effort one could convert

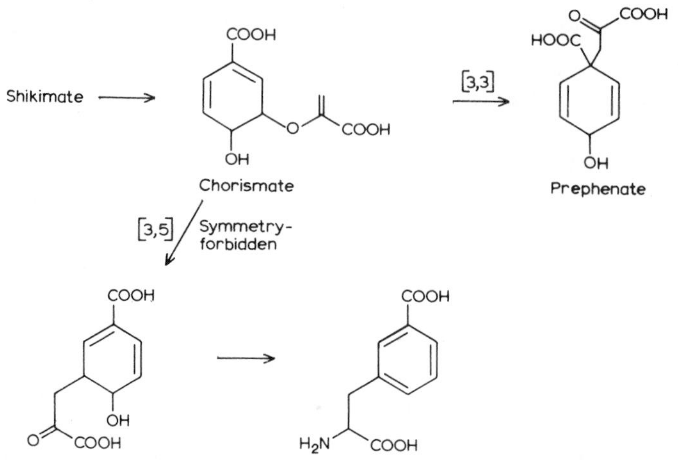

SCHEME 1 (Arigoni). Biosynthesis of *m*-carboxyphenylalanine.

from the (*T, P, m*) system to the (*T, V, c*) system; that is not frequently done, perhaps because the significance is not understood.

Lipscomb: I agree that this could easily be done in model reactions; both temperature (entropy) and pressure (volume) effects should be studied. It is a mistake not to vary the parameters at one's disposal.

Cornforth: Do you know of any enzyme that does not actually participate in a chemical reaction at the active site, in the sense that there is no electron-sharing between substrate and protein?

Lipscomb: Chorismate mutase is a possible candidate but that is probably a matter of local hydrogen-bonding.

Knowles: Could it not be more of a 'Woodward-Hoffman-ase' or an 'oxa-Cope-ase'?

Arigoni: Formation of the unusual amino acid *m*-carboxyphenylalanine from chorismate (Larsen 1977; Larsen *et al.* 1972) can be represented as the outcome of a [3,5]-sigmatropic rearrangement, which is symmetry-forbidden. We take it for granted that the rearrangement to prephenate proceeds in a synchronous manner since the [3,3]-process is symmetry-allowed, but what about the symmetry-forbidden path?

Knowles: That one is 'not obviously concerted'! (See Doering 1976.)

Arigoni: I guess that both go through discrete intermediates.

Lipscomb: So they are not symmetry-concerted. How much does the enzyme speed up this reaction?

Arigoni: The enzyme has not yet been isolated or purified. The data for the [3,5]-sigmatropic rearrangement are not available.

Lipscomb: The rate enhancement factor for the acceleration by chorismate is 2×10^6 (Andrews *et al.* 1973) (see p. 5).

Knowles: If chorismate mutase catalyses an ordinary [3,3]-sigmatropic shift, we can account for about half the rate acceleration in entropy terms alone. If all the -13 e.u. (which is the $\Delta S^{\ddagger}$ for the uncatalysed reaction; Andrews *et al.* 1973) is frozen out when chorismate sits on the active site, there will be an acceleration of about 10^3.

Lipscomb: It is not known whether the enhancement of rate in the enzyme reaction is an effect primarily on the entropy or the energy of the reaction. Calculations may help in such instances when one knows what the transition state looks like from analogues. The transition-state in chorismate mutase is tight and, so, susceptible to an opposing entropy effect (see p. 5). Andrews *et al.* (1973) suggest that the enzyme reduces ΔH^* by 5–9 kcal/mol.

Topping: Fig. 8 lists significant changes in the pK_a values of groups within certain enzymes, due to changes in the local environment of the group. Are these changes readily reversible? In other words, does the pK_a constantly fluctuate as a result of conformational changes in the enzyme? Presumably such behaviour could be exploited in the catalytic process.

Lipscomb: Many of those pK_a values come from n.m.r. studies of enzyme–inhibitor or enzyme–substrate complexes and so refer to particular groups. Whether they fluctuate, I do not know. If these fluctuations are fast on the n.m.r. time scale one would not be able to detect them except possibly as a broadening of the signal.

Williams: Are you saying that the energetics responsible for the pK_a values are confined locally and are not due to interactions spreading into the protein? If so, they cannot be related to the changes in conformation of the protein. The general point is that if one changes the state of protonation of a protein, we need to know how far the energy change spreads in the structure. If it is spread into the structure, one faces an enormously difficult task to design a model.

Lipscomb: I imagine that it is local. For one example, in his work on acetoacetate decarboxylase Westheimer showed that changes of a reporter group and of the lysine to which this group is attached are due to a nearby positive charge and not to a general lowering of the local dielectric constant near the enzyme. This is surely a local effect.

Williams: Great interest focuses on the control of the energetics of groups by the energy of the fold of protein. The pK_a values can reflect much more than the local group energy.

Vallee: Is there any parameter that relates to the hyperreactivity of certain residues apparent on chemical modification?

Lipscomb: The shifts in pK_a values are almost always consistent with the idea that near the protein the dielectric constant is lower than that of water. The shift in pK_a often changes the available concentration of some reactive group. Also, almost all these changes in pK_a favour the less charged form.

Gutfreund: Besides the contribution of the strain of the protein, another point is that probably in no case does one determine the pK of an isolated group but usually a system of groups. This is certainly so for the well-documented cases discussed here: chymotrypsin, alcohol dehydrogenase and lysozyme.

Williams: But I still want to know what happens when the state of ionization of a group changes; is it a local effect or does it spread into the protein? If the energy of ionization is spread into the protein, there are many ways of manipulating catalytic steps which are not a property of the local nature of the attacking group. Protein catalysts may resemble surface catalysts more than small model molecules.

Gutfreund: In lysozyme conformational changes are correlated with changes in pK_a and similarly in liver alcohol dehydrogenase.

Knowles: When a pK_a value is perturbed from its normal value, there must be an equivalent compensating change in free energy, perhaps in the pK_a of another group or in the conformational free energy of the protein. Professor Williams is asking about the relationship between free energy and structure: how floppy is the enzyme, and how *much* of the enzyme accommodates itself in order to balance a local free-energy change such as pK_a shift?

Lipscomb: Would it help if crystallographers determined structures at various pHs, with and without substrate analogues, in order to try to locate these effects?

Williams: Certainly. Let me tell you about the enzyme phospholipase A_2 (collaborative work with Professor de Haas, Utrecht, and A. Aguiar, Oxford). Drenth *et al.* (1976) have determined the crystal structure of the proenzyme but in the active-site region it cannot bear a close relation to the active enzyme in solution since Drenth's work shows a free histidyl group remote from any carboxy group in the active-site groove. The n.m.r. spectrum (at 270 MHz) of this histidine, even at pH 3, shows that it is not protonated. There must be a His $\cdots$ COOH group. (As in chymotrypsin, the two groups can exist as either His $\cdots$ COOH or HisH+ $\cdots$ COO−. N.m.r. spectroscopy will immediately identify which form is present. This particular protonation–deprotonation is important, as it is said to be part of the catalytic step of chymotrypsin.) Now the n.m.r. studies of phospholipase also show that the way in which the proton interacts with the two groups (here imidazole and carboxy) depends on the conformation of the protein as a whole. One can vary the conformation continuously by varying the temperature or pH. As most of the conformations are in fast exchange, the whole protein, including the active site, is in dynamic flux and the position of the active-site proton is of variable reactivity. This raises many difficult problems.

When we talk about mechanisms how accurately do we know (or do we need to know) where a proton is? Can we tell whether a group is an acid or a base at a given pH? Is the pK_a in a protein cooperatively linked to protein dynamics? I want to force this discussion of energetics away from the consideration of strictly local sites (which is the natural result of extending discussions of models).

Lipscomb: One approach to the problem might lie in further work on the synthesis of bacterial enzymes, for example α-lytic proteinase, in which one can introduce isotopes in various places. Introduction of such labels might yield hyperfine structure in the n.m.r. spectra and give local information at selected points in a protein.

Wüthrich: In deriving individual pK_a values from n.m.r. studies we should differentiate between two limiting cases. An ionizable group in the interior of a globular protein may be shielded from the solvent. Since, even in a small molecule like the basic pancreatic trypsin inhibitor, labile interior protons do not exchange at ambient temperature and neutral pH over a period of two to three years (Masson & Wüthrich 1973), interior ionizable groups may not titrate at all unless the structure is opened up. Opening of the conformation of a given protein might be caused, for example, by deprotonation at the NH_2-terminus; by observing the titration shifts of interior ionizable side-chains one might thus effectively be measuring the pK_a of the terminus. On the other hand, an exposed ionizable group may have its pK_a modified by being located near a charged group.

Lipscomb: I agree. Another such example is the set of metal-binding constants. A protein might assemble around the metal, consequently making the constant a measure, in part, of local protein assembly.

Dunitz: You described the reaction cone (Fig. 17) for the approach of the model nucleophile CH_3O^- to formamide. Is your model sensitive enough to pick up energy differences of, say, about 2 kcal/mol between different approach directions within the cone?

Lipscomb: No, it isn't.

Dunitz: We have evidence that the cone must be asymmetric with an energy difference of the order of 2–3 kcal/mol between the optimal approach direction and the axis of the cone. Schweitzer (1977) and G. Procter (unpublished

(1)   (2)   (3)

work, 1977) have synthesized substituted naphthalene derivatives (1), where X = Me$_2$N or MeO and R = OH, OR' or NR'$_2$ and determined their structures by X-ray analysis. Altogether we have studied six compounds of this type. In general, *peri*-substituents on the naphthalene nucleus bow outwards to get away from each other, but in our molecules there is a strong attractive interaction between the nucleophilic heteroatom and the electrophilic carbon atom of the carbonyl, the plane of the R–C=O group being nearly normal to the naphthalene plane. In principle, the bonds to the two substituents could remain straight with 120° bond angles, as in (2) for the methoxy compound: the bonds C(1)–C(=O) and C(8)–O are almost equal in length. This would give an X $\cdots$ C=O approach angle of about 90°. The six molecules we have studied all display a common distortion pattern in which the group containing the electrophile moves outwards and that containing the nucleophile bows inwards, as in (3). The X $\cdots$ C distance is about 2.5–2.6 Å in every case, and the angle α is 5–10° bigger than β. As a result of the bond-angle distortion, the X $\cdots$ C=O approach angle is increased from 90° to about 100°; that is, the nucleophile tends to approach from behind the carbonyl group rather than from normal to it. The same tendency has been observed in other molecules showing strong N $\cdots$ C=O interaction (Bürgi *et al.* 1973). With a bond-angle bending-force constant of about 40 cal/mol deg^2, the energy increase of the bond-angle distortions should be something of the order of 2 kcal/mol. Thus, if the distorted molecule is to correspond to an energy minimum, the X $\cdots$ C=O interaction energy must be at least this amount more favourable for the 100° approach than for the 90° one. In all six molecules the carbonyl carbon atom is somewhat pyramidalized; it is displaced by about 0.05 Å from the plane of its three substituents *towards* the nucleophile. In the dimethylamide (1; R = NMe$_2$, X = OH) the amide nitrogen atom moves about the same distance from the plane of its three substituents but in the opposite direction, i.e. the nitrogen pyramidalizes in such a way that its lone-pair becomes antiperiplanar to the incipient X $\cdots$ C bond.

Lipscomb: Wouldn't the amide nitrogen atom be non-planar anyway?

Dunitz: No, practically not.

Lipscomb: Formamide has a non-planar nitrogen atom.

Dunitz: Formamide was said to be non-planar on the basis of an old microwave analysis (Costain & Dowling 1960) but a more recent analysis (Hirota *et al.* 1974) using additional isotopic species gives a planar equilibrium structure with a rather flat single minimum quartic potential for nitrogen inversion. It is true that many amides observed in the crystalline state are slightly non-planar (Dunitz & Winkler 1975). However, in our dimethylamide it may be more than a coincidence that the displacement of the nitrogen happens to be in the direction that would be expected from the stereoelectronic arguments advanced by Deslongchamps (1975) and applied to chymotrypsin by Bizzozero & Zweifel (1975). If the nitrogen atom of an amide substrate in chymotrypsin does move away from the attacking nucleophile (the oxygen of serine-195), then inversion at this nitrogen has got to occur before the proton required for the next step can be transferred from the histidine. This seems to present a challenge to experimentalists to devise an experiment by which one could prove whether or not the nitrogen does invert.

Lipscomb: In chymotrypsin both the nucleophile and the electrophile have to approach from the same side but in other hydrolytic enzymes such as carboxypeptidase they approach from opposite sides.

Dunitz: In any case, the experimental evidence I adduced shows that the cone cannot be symmetric.

Lipscomb: Our calculations on the approach of methoxide ion from above the plane of formamide indicate an asymmetric approach, easier a bit toward the hydrogen as compared with the oxygen or amino group of formamide (Scheiner *et al.* 1976).

References

ANDREWS, P. R., SMITH, G. D. & YOUNG, I. E. (1973) Transition state stabilization and enzymic catalysis. Kinetic and molecular orbital studies of the rearrangement of chorismate to prephenate. *Biochemistry 12*, 3492-3498

BIZZOZERO, S. A. & ZWEIFEL, B. O. (1975) The importance of the conformation of the tetrahedral intermediate for the α-chymotrypsin-catalyzed hydrolysis of peptide substrates. *FEBS (Fed. Eur. Biochem. Soc.) Lett. 59*, 105-107

BÜRGI, H. B., DUNITZ, J. D. & SHEFTER, E. (1973) Geometrical reaction coordinates. II Nucleophilic addition to a carbonyl group. *J. Am. Chem. Soc. 95*, 5065-5067

COSTAIN, C. C. & DOWLING, J. M. (1960) Microwave spectrum and molecular structure of formamide. *J. Chem. Phys. 32*, 158-165

DESLONGCHAMPS, P. (1975) Stereoelectronic control in the cleavage of tetrahedral intermediates in the hydrolyses of esters and amides. *Tetrahedron 31*, 2463-2490

DOERING, W. E. (1976) The not-obviously-concerted thermal rearrangements, in *Abstracts of The American Chemical Society Meeting* (New York, April 1976), *Organic Chemistry*, abstr. 14

DRENTH, J., ENZING, C. H., KALK, K. H. & VESSIES, J. C. A. (1976) Structure of porcine pancreatic prephospholipase A_2. *Nature (Lond.) 264,* 373-377

DUNITZ, J. D. & WINKLER, F. K. (1975) Amide group deformation in medium-ring lactams. *Acta Crystallogr. B31,* 251-263

HIROTA, E., SUGISAKI, R., NIELSEN, C. J. & SØRENSEN, G. O. (1974) Molecular structure and internal motion of formamide from microwave spectrum. *J. Mol. Spectrosc. 49,* 251-267

LARSEN, P. O. (1967) *m*-Carboxy-substituted aromatic amino acids in plant metabolism. *Biochim. Biophys. Acta 141,* 27

LARSEN, P. O., ONDERKA, D. K. & FLOSS, H. G. (1972) Steric course and rearrangements in the biosynthesis of phenylalanine, tyrosine and 3-(3-carboxyphenyl)alanine from shikimic acid in higher plants. *J. Chem. Soc. Chem. Commun.,* 842

MASSON, A. & WÜTHRICH, K. (1973) Proton magnetic resonance investigation of the conformational properties of the basic pancreatic trypsin inhibitor. *FEBS (Fed. Eur. Biochem. Soc.). Lett. 31,* 114-118

RÜHLMANN, A., KUKLA, D., SCHWAGER, P., BARTELS, K. & HUBER, R. (1973) Structure of the complex formed by bovine trypsin and bovine pancreatic trypsin inhibitor: crystal structure determination and stereochemistry of the contact region. *J. Mol. Biol. 77,* 417-436

SCHEINER, S. & LIPSCOMB, W. N. (1976) Molecular orbital studies of enzyme activity: catalytic mechanism of serine proteinases. *Proc. Natl. Acad. Sci.. U.S.A. 73,* 432-436

SCHEINER, S., LIPSCOMB, W. N. & KLEIER, D. A. (1976) Molecular orbital studies of enzyme activity. 2. Nucleophilic attack on carbonyl systems with comments on orbital steering. *J. Am. Chem. Soc. 98,* 4770-4777

SCHWEITZER, W. B. (1977) *Röntgenkristallographische Untersuchungen über nukleophile-elektrophile Wechselwirkungen an perisubstituierten Naphthalinen*, Doctoral Dissertation (No. 5948), ETH Zürich

WARSHEL, A. & LEVITT, M. (1976) Theoretical studies of enzymic reactions. Dielectric, electrostatic and steric stabilization of the carbonium ion in the reaction of lysozyme. *J. Mol. Biol. 103,* 227-249

Solvents, interfaces
and protein structure

F. M. RICHARDS and T. RICHMOND

Department of Molecular Biophysics and Biochemistry, Yale University, New Haven, Connecticut

Abstract   Mean packing densities in protein interiors are comparable to those of most organic solids but the variations between small regions may be substantial. Packing defects may be related to allowed structural fluctuations. Molecular surface areas can be correlated with free energies of transfer between different solvents. The proportionality factor will depend, in general, on the nature of the solute and both the solvents. The changes in solvent–protein interfacial area on chain folding are large and the implied changes in free energy from this solvent-squeezing effect are correspondingly large. The strong tendency to minimize surface area is reflected in the globular shape of most protein molecules or domains in larger structures. The formation of isolated units of secondary structure from an extended chain represents about one half of the eventual total area change. The tendencies of amino acids to form β-sheets correlate well with the rank-ordered list based on non-polar area change for each residue type. The calculated area changes for helix and sheet formation are not identical in rank order. The rank-ordered list for α-helix formation correlates satisfactorily with the probability list prepared from actual structures if glutamic acid and tyrosine are removed. What special characteristics unrelated to surface area these two amino acids might have is not clear. Tertiary structure formation from preformed secondary structural units can be rank ordered on area change and possible nucleation sites can be identified. A prediction scheme for helix–helix interactions is proposed. The hydrophobic force begins to be felt when two helices are about 0.6 nm (6 Å) from their final contact positions. Interfacial surface tension is a logical parameter to relate free energy and solvent contact area, but this macroscopic parameter must be used with great caution. It is suggested that water in the deep grooves, characteristic of the active sites of many enzymes, may have a substantially higher fugacity than bulk water as indicated, at least qualitatively, by the Kelvin equation based on surface curvature. Such water would be more easily displaced than its plane surface counterpart and could contribute significantly to ligand–binding energy. This factor would be in addition to the usual solvent entropic effects associated with surface area reduction on association.

Most proteins whose structure is known in detail are globular with low axial

23

ratios. Proteins such as the immunoglobulins which are both extended and flexible have a domain structure where the individual domains are compact, globular, and of low axial ratio. The fact that highly-asymmetric peptide chains with a variety of different sequences tend to form such compact structures indicates the existence of a large driving force tending to reduce the interfacial area between the peptide and the solvent. In the terminology of polymer chemistry, water must be considered a poor solvent for long peptides. The interface between the solvent is a critical region both for the native structure of the protein and for its interaction with ligands of all sizes.

PACKING OF PROTEIN ATOMS

The compact structures that are the result of the folding process are essentially free of solvent and without large voids. The packing density is equivalent, on the average, to that found for most organic compounds in the crystalline state (Richards 1974). Finney (1975) and Chothia (1975) have noted that the mean packing density for the various types of amino acid residues is close to that of the crystalline amino acids and that the standard deviations are of the order of 10% of the mean residue volume. Solely on the basis of volume and with no regard for shape, none of the larger residues could be replaced, in an otherwise fixed structure, by a group which differed by more than the equivalent of a single methylene group. The smaller residues are even more closely circumscribed. Severe constraints on the final structure thus may be imposed by packing alone, before any consideration is given to the special requirements of hydrogen-bonding groups or of those with formal charges. The flexibility of the structures undoubtedly varies with position, but local motion may frequently allow some covalent, structural changes without severe energy deficits and thus render the packing criteria less stringent than implied by the comments above (Gelin & Karplus 1975). Even in the static case, packing variations over small regions may be large (Richards 1974). Use of packing criteria in acceptance or rejection of trial structures required averaging over substantial volumes of space. Uncertainty over proper procedure in the ill-defined solvent-contact interface and time-consuming computational algorithms have slowed application of these criteria in any detailed way to general folding studies. However, the packing defects may be intrinsically interesting and related to principal modes of motion.

MOLECULAR AREAS AND SPECIFIC FREE ENERGIES OF TRANSFER

Various algorithms exist for the calculation of the surface area of a molecule. The differences in technique of calculation are probably not important at this time, but the surface whose area is to be measured must be clearly defined (Richards 1977). The 'accessible surface' of Lee & Richards (1971) and the 'cavity' surface of Hermann (1972) refer to the locus of the centre of a spherical probe touching the van der Waals envelope of a molecule. The 'contact surface' refers to that part of the van der Waals surface in contact with the probe. The 'molecular surface' includes the disconnected patches of the contact surface and the reentrant sections connecting them to produce a continuous sheet. Although closely related these measures of surface area are numerically different (see Richards 1977). When two conformations of a flexible molecule are known in detail, the surface area of each can be calculated and the area difference estimated accurately.

For a homologous series of compounds the free energy of transfer between two solvents has been found to be linearly related to surface area (Hermann 1972; Chothia 1974; Harris *et al.* 1973; Reynolds *et al.* 1974). The process of folding a peptide chain may be considered as equivalent to transferring residues of the chain from an aqueous solvent to an environment with different characteristics. The change in area of contact with water between the unfolded and folded states can be used as an estimate of the free-energy change associated with this 'solvent-squeezing' component of the overall process. In order to derive numerical values, one must know the proportionality constant relating area change and free-energy change. There is no disagreement between investigators in this field that the effect is large and is perhaps the principal driving force for the formation of the compact native structures. However, there is considerable disagreement on what the correct values are and even on the approach to deriving them.

The most commonly used reference data are transfer free energies based on partition or solubility experiments. The appropriate areas must now be decided. Both Hermann (1977) and Gelles & Klapper (1978) have pointed out that the area of a flexible molecule varies with conformation. The appropriate reference area should be the weighted mean of the various conformations where the Boltzmann factor includes not only the intramolecular torsional potentials but area difference as well. The factor relating transfer free energy to area will clearly depend directly on the reference areas chosen. Fortunately, the effects are not large for relatively small molecules. For the normal alkanes up to octane the area difference between the extended and most folded conformations is less than 10% of the total area. No amino acid side-chain has more than five rotatable bonds and most have only two. Thus the problem of starting conformations can probably be set aside at this time in view of the more serious difficulties of other kinds.

Although amino acids dissolved in water may be an appropriate reference state for an extended peptide chain, the organic solvent to which the transfer should be made is much less clear. The interior of a protein is more like a solid than a liquid in that the individual atoms can be uniquely located by diffraction procedures. However, it is also true that the apparent temperature factors, representing mean square displacement, are much larger than those found in crystals of small organic molecules at the same temperature. The appropriate reference state for free-energy estimates is thus unlikely to be either solution in an organic solvent or an organic crystalline phase. Nevertheless, Gelles & Klapper (1978) have concluded from an examination of fatty acid solubilities that the packing density of the organic phase has no important effect on the transfer free energies and thus the solution partitioning values may be safely used. (The significance of the secondary phase transition in fatty acids, which renders the crystals partially liquid-like well below their melting temperature, should perhaps be considered further.)

Apart from the solid or liquid nature of the protein interior its polarity is the next consideration in deciding on reference organic solvents. The interior is a mixture of polar and non-polar parts non-uniformly distributed. Internal contact areas have been examined in detail by J. L. Finney (personal communication). The environment of different individual residues of the same class can be very different and thus the use of a single reference solvent may be untenable (Nandi 1976; Gelles & Klapper 1978). Specific free energies of transfer for a given compound may differ by more than 40% depending on the specific organic solvent used. Even within the so-called non-polar side-chain category it is not clear that for a given solvent pair one should expect the specific transfer free energy of alkanes to be the same as that of aromatic compounds. The fact that in certain systems all-non-polar side-chains fall on the same straight line is probably fortuitous. Referred to accessible surface areas, specific transfer free energies between 20 and 33 cal $Å^{-2}$ mol^{-1} have been suggested (Hermann 1977; Richards 1977). These values correspond to about $70 - 115$ cal $Å^{-2}$ mol^{-1} based on contact areas. The specific transfer free energies that should be used remain a subject for future research. For the rest of this paper emphasis will be placed on areas with the assumption that, in due time, they can be appropriately converted into free energies when required.

AREA CHANGES ON CHAIN FOLDING

Even for small proteins the change between the extended chain and the native structure corresponds to a reduction in area of about two-thirds. The folding process may be divided up hypothetically into the formation of isolated units of secondary structure followed by the assembly of these units into the final tertiary structure. The total area change is divided about equally between these two steps. The 'solvent-squeezing' effect should thus be contributing significantly to the formation of secondary structure. This force is non-specific in the sense that any process leading to area reduction will be encouraged.

Secondary structure

The contact areas for residues in standard secondary structures are listed in Tables 1 and 2. The reference chains are poly(alanine) with a single guest residue inserted in the middle. In going from an extended chain to an anti-parallel β-sheet the change in mean contact area per residue is $20.5 \pm 3.3 \text{Å}^2$; from extended chain to α-helix, the change is $15.8 \pm 2.1 \text{ Å}^2$. The difference between these standard areas for an α-helix and the mean residue areas by class calculated for the actual helices in myoglobin (considered one at a time in isolation) is $3.8 \pm 3.4 \text{ Å}^2$. The effect of the larger residues in actual helices is thus apparent but is not large, and the conclusions based on an examination of the standard values are likely to be qualitatively correct.

The area changes based on the main-chain α-carbon atom and the side-chain carbon and sulphur atoms were summed for each residue type. The residues were rank-ordered on these changes giving two lists, one based on the coil-to-α-helix values and the other on the coil-to-β-sheet values. These lists were then compared with α-helix or β-sheet probability lists, such as those provided by Chou & Fasman (1977), determined from known protein structures. The rank correlation coefficients (Siegel 1956) are given in Table 3. The area changes are plotted against residue type in Fig. 1. The total area changes, polar and non-polar, correspond to about 35–40% of the extended chain value for each residue type. The high rank correlation between the poly(alanine) helix data and the actual myoglobin helix data (Table 3, line 1) shows that the larger area change for actual helices, mentioned earlier, is uniformly distributed and does not significantly change the residue rank order.

The high correlation between C→β and P_β (Table 3, line 4) seems to indicate that consideration of loss of non-polar contact area only is sufficient to predict the β-sheet-forming tendency of an amino acid. Although obviously related, the area changes for C→α are not the same as for C→β and the rank orders are somewhat different. (This is reflected in the fluctuations seen in Fig. 1.) However, the correlation of C→α with P_α, although significant, is low. The removal of only two amino acids from the list (Tyr and Glu) raises the correlation coefficient of the residual list to a high value (see Table 3, line 7).

TABLE I

Residue contact areas in standard structures[a] (Å²)

Amino acid residue	Extended chain[b] $Ala_3 \cdot X \cdot Ala_3$			α-Helix[b] $Ala_4 \cdot X \cdot Ala_4$			Antiparallel β-sheet[b] $Ala_7/Ala_3 \cdot X \cdot Ala_3/Ala_7$			Myoglobin α-helices[c] mean values		
	Side-chain		All atoms[d]	Side-chain		All atoms[d]	Side-chain		All atoms[d]	Side-chain		All atoms[d]
	Non-polar	Polar		Non-polar	Polar		Non-polar	Polar		Non-polar	Polar	
Ala	22.1		33.9	15.9		20.0	15.6		16.7	13.4		16.9
Arg	23.1	37.7	72.6	17.7	34.0	54.8	14.8	38.1	53.8	19.6	29.0	49.8
Asn	12.7	17.6	42.0	9.0	16.5	27.7	7.6	12.7	21.3	3.7	19.9	23.9
Asp	13.6	14.7	40.2	9.7	13.1	25.6	7.6	12.1	20.7	6.5	11.8	19.5
Cys	28.7		40.6	22.3		25.4	20.6		21.5			
Gln	14.6	25.8	52.2	12.1	20.8	35.9	7.1	24.8	32.8	6.9	26.9	36.0
Glu	17.2	19.2	48.2	13.6	16.1	32.7	9.0	19.3	29.1	9.7	13.2	24.5
Gly			24.5			12.9			10.4			10.5
His	28.8	13.7	53.7	25.0	9.6	37.0	18.0	12.4	30.9	26.6	11.6	40.4
Ile	42.7		53.7	37.5		39.0	30.6		31.4	30.8		31.9
Leu	39.6		51.0	31.7		34.7	31.2		31.5	31.8		33.1
Lys	32.8	18.1	62.8	25.7	16.9	45.6	24.5	18.4	43.7	26.8	13.6	43.9
Met	48.9		60.6	39.7		42.7	39.5		40.4	37.5		38.9
Phe	52.7		63.7	43.6		45.8	39.7		40.1	37.1		39.3
Pro	33.2		41.0	16.9		21.5	21.2		21.3	22.5		23.7
Ser	14.1	7.6	33.9	10.5	6.2	20.4	9.7	5.9	16.7	8.2	4.3	13.7
Thr	23.0	6.8	41.2	20.9	5.3	28.3	13.2	5.3	19.5	17.0	5.6	26.3
Trp	65.8	3.4	79.7	57.8	1.5	60.9	45.2	4.1	49.8			
Tyr	44.7	8.4	64.0	36.1	8.0	46.4	30.7	8.9	39.9	45.1	7.8	52.9
Val	35.6		46.7	31.2		32.7	22.8		23.7	24.4		30.6
Mean values	31.3	15.7	50.3	25.1	13.4	34.5	21.5	14.7	29.8	21.6	14.4	25.5

[a] The difference in definition between contact areas and accessible or cavity areas should be noted. See text and Richards (1977).

[b] For all residues $\chi_1 = -60°$, $\chi_2\text{-}\chi_n = 180°$ except for rings where $\chi_2 = 90°$. For main chain, $\omega = 180°$ in all cases. Extended chain: $\varphi = -139°$, $\psi = -135°$; α-helix: $\varphi = -58°$, $\psi = -48°$; antiparallel β-sheet: $\varphi = -120°$, $\psi = -140°$ (Arnott & Dover 1967; Arnott et al. 1967; Chothia 1973).

[c] Residues in first and last turn of each helix were deleted in computing the averages.

[d] Total of main-chain and side-chain atoms, polar and non-polar.

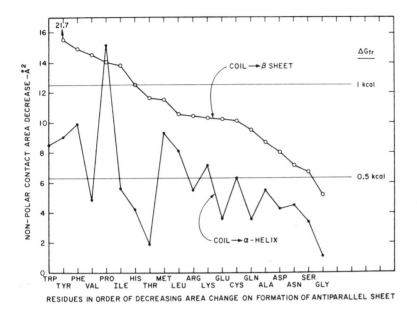

FIG 1. Area changes on going from an extended chain (coil) to β-sheet or to α-helix are shown for each residue type. The α-helix data were calculated for the peptide $Ala_4 \cdot X \cdot Ala_4$. The β-sheet data were calculated for a twisted, three-strand, antiparallel sheet $Ala_7/Ala_3 \cdot X \cdot Ala_3/Ala_7$. The residues on the abscissa are listed in the order of decreasing area change from left to right for the β-sheet data. The β-sheet order corresponds closely to the P_β list of Chou & Fasman (1977), indicating the probability of finding a residue in a β-sheet from analysis of known structures. The lack of correlation between the helix and sheets lists is reflected in the fluctuations of the helix line. The free energy of transfer lines is based on a proportionality constant of 80 cal $Å^{-2}$ mol^{-1} (referring to contact area not cavity area). The energy values should only be taken as a rough indication of magnitude.

Again loss of non-polar contact area appears sufficient to predict the order of this residual list. The behaviour of Tyr and Glu is anomalous and must be related to some property other than surface area. What distinguishes these amino acids from others is not clear.

The non-polar area changes for β-sheet formation are larger than for the α-helix. The mean area change is 11.7 $Å^2$/residue for the β-sheet and 6.4 $Å^2$/residue for the helix or about 0.9 and 0.5 kcal/residue, respectively. The estimated difference in stability of 0.4 kcal/residue is large and for chains of any length renders the α-helix an improbable secondary structure. Although the β-sheet may be a more common, general, structural feature in protein, clearly other effects must enter at the secondary or higher structural level to reduce this energy difference between the two conformations. Alternatively, one might imagine that the initial nucleation sites are β-sheet regions with the

TABLE 2

Contact areas for main-chain atoms in standard structures (Å^2)

	N	C	O	Cα	All
Extended chain					
18 residue mean	2.0±0.1	0.3±0.1	7.4±0.1	1.9±0.3	11.6
Gly	3.5	1.0	8.0	12.0	24.5
Pro	0.0	0.5	5.3	2.0	7.8
α-Helix					
18 residue mean	0.0	0.2±0.1	0.7±0.3	1.7±0.6	2.6
Gly	0.2	0.9	0.9	10.9	12.9
Pro	0.1	0.4	0.9	3.2	4.6
β-Sheet					
18 residue mean	0.7±0.3	0.0	0.1	0.0	0.8
Gly	2.4	0.7	0.5	6.8	10.4
Pro	0.0	0.0	0.1	0.0	0.1
Myoglobin α-helices[a]					
Residue mean[b]	0.2±0.3	0.1±0.2	0.6±1.2	1.3±1.5	2.2
Gly mean	0.1	0.8	0.3	9.3	10.5
Pro	0.0	0.4	0.6	0.1	1.1

[a] Residues in first and last turn of each helix were deleted in computing the averages.
[b] Main chain atoms of all residues except Gly and Pro.

associated helices laid down subsequently. However, the majority of synthetic poly(amino acids), whose conformational transitions have been studied, appear to form isolated helices rather than sheets.

Tertiary structure

Starting with preformed secondary structural units, formation of the tertiary structure involves additional loss of area. The packing of the helices in myoglobin was examined a few years ago by Ptitsyn & Rashin (1973) and more recently by Richmond & Richards (1978). The interlocking of the side-chains appears to prevent extensive motion of either translation along or rotation about the axis of any of the helices (Fig. 2). If one starts with the preformed helices, the principal folding path appears to involve translations perpendicular to the helix axes as the helices approach each other.

Solvent exclusion, and thus the appearance of the hydrophobic force encouraging association, begins about 6 Å away from the final position where the helices are packed as in the native structure. The area change is roughly linear between 6 Å and this final position. The helix pairs can be ranked on the

TABLE 3

Rank list correlations based on area changes for non-polar atoms

C→α:	Residue list ranked on area change during conversion of extended chain into α-helix for $Ala_4 \cdot X \cdot Ala_4$.
C→β:	Residue list ranked on area change during insertion of extended chain into centre of three-strand antiparallel sheet based on $Ala_3 \cdot X \cdot Ala_3$.
Mb(C→α):	Residue list ranked on area changes for myoglobin sequences during conversion of hypothetical extended chain into actual α-helices.
P_α or P_β:	Residues ranked according to probability of occurrence in α-helices or β-sheets from survey of known proteins (Chou & Fasman 1977).

List 1	List 2	Spearman rank correlation coefficient, r_s	r_s values for	
			$P=0.05$	$P=0.01$
19 Amino acids (Pro omitted)				
1 C→α	Mb (C→α)	0.96	0.39	0.55
2 C→α	P_α	0.39		
3 C→α	P_β	0.50		
4 C→β	P_β	0.74		
5 C→β	P_α	0.22		
6 P_α	P_β	−0.06		
17 Amino acids (Pro, Glu, Tyr omitted)				
7 C→α	P_α	0.65	0.41	0.58
8 C→α	P_β	0.43		
9 C→β	P_β	0.76		
10 C→β	P_α	0.41		
11 P_α	P_β	0.25		

magnitude of the final area change. If the minimum final distance between helix axes is greater than 11–12 Å, the area change is small and thus also the corresponding free-energy change. For tighter contacts, however, the area changes can be large and correspond to tens of kilocalories in free energy. These tight contacts are good candidates for nucleation sites in tertiary folding.

A prediction scheme has been suggested for identifying possible strong helix–helix interaction sites based on a set of standard area changes for each type of residue (Richmond & Richards 1978). Helices idealized as cylinders formed from sheets of close packed spheres are shown in Fig. 3. In the α-helix ($m = 4$), each residue is surrounded by six nearest, but non-equivalent, neighbours. Should such a patch form part of a helix–helix interaction site, area changes can be predicted from a known amino acid sequence. The residue i is referred to as the central residue of a patch. The nature of the helix interaction (i.e., separation of the helix axes and the interaxial angle) depends

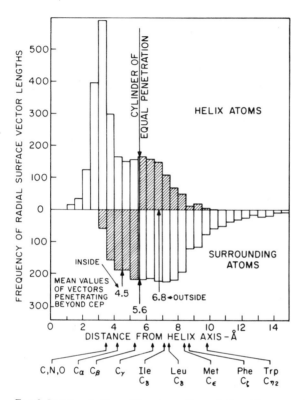

FIG. 2. Interpenetration of the atoms of an α-helix with surrounding atoms in the structure. The data are taken from the helices in myoglobin. Radial vectors from helix axes terminate at the atom surface. The frequency of vector lengths is shown. The shaded portions have equal area and define the radius of the cylinder of equal penetration. The mean atom overlap is 2.3 Å although much larger overlaps occur in particular regions. The positions of the outer surface of some selected atoms for helix main chain and fully extended side-chain atoms are also shown. (Reprinted, with permission, from Richmond & Richards 1978.)

on the size of this central residue. A glycine in this central position allows tight contacts with interaxial angles of about 80° — Class I. With Ala, Val, Ile, Ser, Thr or Cys as central residue, the separation is slightly larger and the angles smaller and more variable — Class II. With yet larger central residues only parallel association of helices (angles <20°) provides an interaction of significant strength — Class III. Predicted contact areas for myoglobin as a function of sequence position are shown in Fig. 4. If we accept relatively high area (and thus energy) cut-off values, the number of possible strong interaction sites is small. Use of such an algorithm might substantially limit the number of helix–helix interactions which would have to be listed in a general

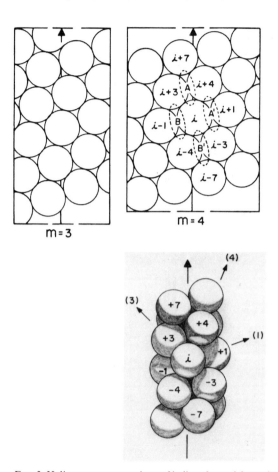

FIG. 3. Helix net representations of helices formed from sheets of close packed spheres. The index number m identifies each possible helix. With $m = 2$ the helix corresponds to a fully extended chain; $m = 3$ corresponds closely to a 3_{10} helix and $m = 4$ to an α-helix. A pictorial representation of the $m = 4$ helix is shown below. The geometrical difference between the ± 4, ± 3, and ± 1 positions can be easily seen. This is obscured in the planar representation. For helix–helix interactions the potential for an interaction site is initially estimated from the possible area contributions of a central residue i and its six immediate neighbours. (Reprinted with permission from Richmond & Richards 1978.)

folding algorithm. These considerations have not yet been extended in detail to the more common helix–sheet types of interaction.

SURFACE TENSION

The correlations between free energy of transfer and molecular surface area lead naturally to a consideration of surface tension as the connecting

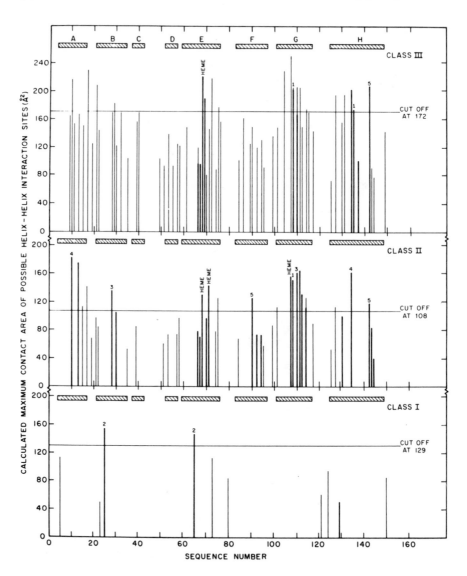

FIG. 4. Prediction of helix–helix interaction sites in myoglobin, for all possible central residues by class. See the text for the list of possible central residues in each class. The predicted maximum area change for any site is shown on the ordinate. For the thick vertical lines all neighbouring residues are in an actual helix. For the thin lines one or more residues used in the summation are not in an actual helix. The lettered bars cross-hatched at the top show the positions of the helices in the myoglobin structure. The numbers over certain lines refer to the actual sites found in the structure. Each number appears twice, once for each helix involved in the interaction site. The haem–helix contacts are also shown. Possible strong sites are considered to be those extending above the indicated cutoff values. The details of the procedure are discussed by Richmond & Richards (1978) from which this figure is reprinted, with permission.

parameter. Detailed calculations on surface tension and cavity effects were made some years ago by Sinanoglu and applied to nucleic-acid structures (Sinanoglu & Abdulnur 1965; Sinanoglu 1968). Melander & Horváth (1977) have extended this approach to hydrophobic interactions in both the precipitation and chromatography of proteins. Excellent linear correlations were obtained for salting-out constants and molal surface-tension increments for several salts. The latter property was suggested as the basis for a natural lyotropic series.

The water–hexane interfacial tension is about 51 mN/m (dyne/cm). For a change in area this corresponds to a free energy of about 75 cal $Å^{-2}$ mol^{-1}. For the benzene–water interface the number is about 50 cal $Å^{-2}$ mol^{-1}. These numbers are the right order of magnitude but surface tension is a macroscopic concept and the relationship between the area of the macroscopic plane surface and any of the molecular surface representations is not entirely clear. If all atoms of the molecule and the probe molecule are the same size and arranged as close-packed spheres, then the contact surface area would be 0.55 times that of the plane surface. The free energy would thus be 90–135 cal $Å^{-2}$ mol^{-1} of contact surface area. These numbers bracket the range listed by Gelles & Klapper (1978) for amino derivatives in a series of solvents but the agreement may be fortuitous. The interfacial tension between octanol and water is only 8.5 mN/m. This number clearly does not fit the measured free energies of transfer to this solvent. The orientation of octanol molecules at the interface makes this region different from the bulk solvent as with any surface active molecule. The free energies of transfer will depend on some bulk property such as the dielectric constant suggested by Gelles & Klapper (1978).

SURFACE CURVATURE

Whatever the actual interfacial tension, the Kelvin relation for the vapour pressure of water as a function of curvature should apply. Although it is true that the usual form of the equation may not be exact with radii of molecular dimensions, the trend should be correct. The attempt to minimize solvent contact will tend to produce particles with as smooth a surface as possible. However, the actual result frequently leaves a deep groove in the protein surface which may have an inner radius of curvature of the order of 5 Å or even less. Based on the Kelvin equation this would give an increase in vapour pressure of water corresponding to more than 1 kcal/mol. This would be a destabilization energy for the water in the groove and implies that this solvent should be easier to displace than bulk water. This energy differential would appear in the binding constant for a ligand and should be a significant contribution even if only a few such water molecules are displaced. Active-site grooves are normally thought to provide steric constraints on the ligands to be bound in addition to the correct geometry for polar group interactions. It is hard to see how steric constraints on a substrate lead to a large binding free energy. Entropic considerations would lead to the reverse conclusion. Release of solvent and the entropy increase commonly ascribed to that reaction would occur to a substantial extent on the association of the ligand with any part of the protein surface. It seems that the energetics of water in deep grooves might be an equal partner with the requirements for specific interaction, steric fit, and surface-area reduction in developing specificity in ligand binding.

The convex surfaces of proteins generally have much larger curvatures than the deep grooves. The effects on water fugacity will be of opposite sign but also numerically much smaller.

These surfaces tend to be rough on a molecular scale (Richards 1977). The number of water molecules that it takes to form a coherent monolayer on a protein of molecular weight 20 000–30 000 corresponds to about 0.6 g H_2O/g of protein or about twice the normal estimates of 'hydration'. Water of hydration is recognized as solvent which is somehow different from bulk water by some particular physical technique. The corollary would be that solvent not so recognized must be indistinguishable from bulk water, the reference material. If half the water in direct contact with the protein surface is normal, this puts the dividing line between the water and the protein and not outside the first solvent monolayer. Such considerations would be especially relevant to hydrodynamic phenomena. In general, it is not clear that the set of water molecules specified as abnormal by one technique are necessarily the same as those affecting a different technique. Were it possible to tag all the hydration molecules identified by any method, the total number might be much larger than 0.3 g/g and might indeed provide the monolayer that is geometrically required.

References

ARNOTT, S. & DOVER, S. D. (1967) Refinement of bond angles of an α-helix. *J. Mol. Biol. 30*, 209-212

ARNOTT, S., DOVER, S. D. & ELLIOTT, A. (1967) Structure of β-poly-L-alanine: refined atomic coordinates for an antiparallel β-pleated sheet. *J. Mol. Biol. 30*, 201-208

CHOTHIA, C. (1973) Conformation of twisted β-pleated sheets in proteins. *J. Mol. Biol. 75*, 295-302

CHOTHIA, C. (1974) Hydrophobic bonding and accessible surface area in proteins. *Nature (Lond.) 248*, 338-339

CHOTHIA, C. (1975) Structural invariants in protein folding. *Nature (Lond.) 254*, 304-308

CHOU, P. Y. & FASMAN, G. D. (1977) *Adv. Enzymol. 45*, in press

FINNEY, J. L. (1975) Volume occupation, environment and accessibility in proteins. The problem of the protein surface. *J. Mol. Biol. 96,* 721-732

GELLES, J. & KLAPPER, M. H. (1978) Pseudo-dynamic contact surface areas: estimation of apolar bonding. *Biochim. Biophys. Acta 533,* 465-477

GELIN, B. R. & KARPLUS, M. (1975) Sidechain torsional potentials and motion of amino acids in proteins: bovine pancreatic trypsin inhibitor. *Proc. Natl. Acad. Sci. U.S.A. 72,* 2002-2006

HARRIS, M. J., HIGUCHI, T. & RYTTING, J. H. (1973) Thermodynamic group contributions from ion pair extraction equilibria for use in the prediction of partition coefficients. Correlation of surface area with group contributions. *J. Phys. Chem. 77,* 2694-2703

HERMANN, R. B. (1972) Theory of hydrophobic bonding. II. The correlation of hydrocarbon solubility in water with solvent cavity surface area. *J. Phys. Chem. 76,* 2754-2759

HERMANN, R. B. (1977) Use of solvent cavity area and number of packed solvent molecules around a solute in regard to hydrocarbon solubilities and hydrophobic interactions. *Proc. Natl. Acad. Sci. U.S.A. 74,* 4144-4145

LEE, B. & RICHARDS, F. M. (1971) The interpretation of protein structures: estimation of static accessibility. *J. Mol. Biol. 55,* 379-400

MELANDER, W. & HORVÁTH, C. (1977) Salt effects on hydrophobic interactions in precipitation and chromatography of proteins: an interpretation of the lyotropic series. *Arch. Biochem. Biophys. 183,* 200-215

NANDI, P. K. (1976) Thermodynamic parameters of transfer of *N*-acetyl ethyl esters of different amino acids from organic solvents to water. *Int. J. Peptide Protein Res. 8,* 253-264

PTITSYN, O. B. & RASHIN, A. A. (1973) Samoorganizatsiia molekuly mioglobina. *Dokl. Akad. Nauk S.S.S.R. 213,* 473-475

REYNOLDS, J. A., GILBERT, D. B. & TANFORD, C. (1974) Empirical correlation between hydrophobic free energy and aqueous cavity surface area. *Proc. Natl. Acad. Sci. U.S.A. 71,* 2925-2927

RICHARDS, F. M. (1974) The interpretation of protein structures: total volume, group volume distributions and packing density. *J. Mol. Biol. 82,* 1-14

RICHARDS, F. M. (1977) Areas, volumes, packing and protein structure. *Annu. Rev. Biophys. Bioeng. 6,* 151-176

RICHMOND, T. & RICHARDS, F. M. (1978) Packing of α-helices: geometrical constraints and contact areas. *J. Mol. Biol. 119,* 537-555

SIEGEL, S. (1956) *Nonparametric Statistics for the Behavioral Sciences*, pp. 202 *et seq.*, McGraw-Hill, New York

SINANOGLU, O. (1968) Solvent effects on molecular associations, in *Molecular Associates in Biology* (Pullman, B., ed.), pp. 427-445, Academic Press, New York

SINANOGLU, O. & ABDULNER, S. (1965) Effect of water and other solvents on the structure of biopolymers. *Fed. Proc. 24,* part III, S-12-S-23

Discussion

Franks: Is the linear change in area as the helices come closer than 6 Å (see p. 30) expressed as a function of potential energy or of a change in the force between the two helices?

Richards: The area change is a purely geometrical calculation. By implication this change in area is directly proportional to a change in free energy.

Franks: Is this force an attraction or a repulsion?

Richards: The decrease in surface area results in a net attractive force. There is no repulsion until the atoms of the helices are in van der Waals' contact.

Lipscomb: Are the axes of the helices perpendicular as they approach?

Richards: So far we have only put them in the correct angular position and let them come together along the line of closest approach.

Franks: With 6 Å as a cut-off, are you assuming that there is no attraction beyond the first layer of water molecules?

Richards: Side-chains, on average, are not long enough to have any significant interactions when they are further apart than 6 Å. Remember, we are considering the physical exclusion of the solvent as the sole effect in these estimates.

Franks: In experiments and calculations with model compounds we have found that the attraction between apolar groups persists beyond 6 Å, even to 10–12 Å (Franks *et al.* 1976; Clark *et al.* 1977). What the origin of the hydrophobic interaction might be, I do not know. We call it hydrophobic because it has all the symptoms of the hydrophobic interaction; that is, it is driven by entropy, associated with a large change in heat capacity and its magnitude varies monotonically with the size of the alkyl groups involved (Franks 1975).

Richards: How do you calculate it?

Franks: From osmotic second virial coefficients and a simple model for the hydration shell we have calculated a possible form of the curve describing the potential of mean force as a function of separation between two alkyl groups in water. The net attraction is a long-range effect. That is why I wonder if the distance of 6 Å is an appropriate choice for a cut-off.

Richards: It is an arbitrary choice in that only geometrical factors have been taken into account.

Vallee: Are you assuming that there is no diffusion limitation on account of the size of the crystal — that is, water and other solutes can exchange freely without any restriction because of pore size?

Richards: The restriction of pore size has to be taken into account. Some of these solutes are big. The crystal contains about 20 Å unit pores and the solute diameters can be at least 10–15 Å.

Vallee: What should the size of the crystal be in order to have no diffusion limitation at all?

Richards: These are big crystals that we are measuring.

Blundell: You are taking the coordinates of the atoms in protein crystals as fixed points in the crystal. Refinements of atomic coordinates of a protein derived by X-ray analysis have indicated that residues on the surface are only occupying the positions initially defined for them for a fraction of the time; the occupancies of those positions are low. Most of the water molecules also have low occupancy. If one allowed for the fact that the side-chains on the surface were moving around, would they not be more accessible than your calculations indicate for a good proportion of the time? This movement might vary the relative accessibility of the hydrophilic side-chains with regard to the hydrophobic ones.

Richards: It is true that some side-chains are sufficiently poorly defined that one does not see them. One cannot see them as well as one can see the water sometimes. This change in accessibility is a real problem. We have not dealt yet with the flexibility of side-chains, simply for computational reasons. I imagine that the average accessibility will not change when we do a proper time average, even though that of individual residues may change.

Blundell: A large polar side-chain might occupy several different orientations for, say, 70% of the time and the rest of the time be closer to a group with a complementary charge, in which arrangement it would be comparatively inaccessible.

Richards: True; but a change in one residue will be compensated by changes in others. If one is focusing on a specific residue then one has to worry about time-dependent shifts and the relative occupation of the different conformational states.

Lipscomb: In the active site of carboxypeptidase A we see water molecules at the expected electron density for full occupancy. Most of the solvent molecules (we see only 80% of the water molecules) are much lower in electron density than they would be if they had full occupancy. Moreover we see many maxima in electron density which are far too close together to be in anything but averages of statistical occupancy in different unit cells. The structure, I believe, is not like ice but more like liquid. Movement of small substrate molecules is apparently not limited by diffusion in these large crystals. The fact that we do not see the remaining 30% of the water molecules is consistent with a disordered water-like structure rather than an ordered ice-like structure.

Richards: Are you surprised that you see as much as you do?

Lipscomb: Yes.

Knowles: Are you saying that there is no conflict between the statement that the water you see in carboxypeptidase A crystals is more like bulk water and yet many water molecules show full site occupancy? Is this because diffusion occurs by a matrix-site hopping process rather than by a more ill-defined slithering?

Franks: To a diffusing species (be it water or another small molecule) the hydrated crystal resembles bulk water and, just like water in a carbohydrate gel, it diffuses as rapidly as it does in bulk water.

Richards: How does this fit in with Professor Knowles' question (which echoes my confusion, too)? The process of diffusion in not-quite-crystalline solids may be difficult to assess when one looks merely at a bulk diffusion coefficient. The diffusion may not be all that retarded even though the solvent is localized to a substantial extent.

Franks: In any hydrated material which, mechanically, behaves as a solid, water will appear to be localized but, on a microscopic scale, fast exchange and fast diffusion persist (Packer 1977). I sometimes explain the terms 'structure' and 'structure in the liquid' to students as follows: imagine taking periodic snapshots of a dinner party in a restaurant with four people at each table. Pictures taken throughout the evening will show the same four people sitting round each table; that is analogous to a solid. In the analogy of liquid, the 'sit-down' dinner is replaced by a buffet supper with enough room at each table for four people to fill their plates. The pictures would still show four people at each table, but each picture would show four different people. Structurally, if the people are as indistinguishable as water molecules, the result is the same, but dynamically it is not, because at the buffet supper people swap tables.

Wüthrich: In all, protein crystals contain about 40% water. Professor Lipscomb just said that one can detect water molecules which are somewhat closer to the protein surface than the average for interspatial water. Can your measurements distinguish between those localized water molecules and the bulk of the interspatial water?

Richards: Not by diffusion measurements. I suppose one could distinguish them, in principle, by the equilibrium distributions of solutes. Calculations on the real pore structures are difficult. However, imagine the pore in the protein as a cylinder with diameter d. If a layer of water is immobilized on the surface of the protein (as well it might be), the diameter of the pore is reduced by 6 Å and that, as a large effect in pore restriction, should show up. The correlation with the diffusion coeffficients and large solutes which are comparable in size to the pore indicates that d is a better diameter to use than $(d - 6)$. That implies that the water immediately adjacent to the protein surface is exchanging fast enough to appear to be bulk water as far as the solute molecule is concerned.

Wüthrich: Is it correct to say that your measurements give information about the crystals rather than the molecular structure of the enzyme?

Richards: That is correct.

Williams: In your paper you were referring to globular proteins and not to proteins in general. Is that right?

Richards: Nothing I have said was based on any evidence other than that from globular proteins!

Williams: Many proteins do not conform to your picture of a protein staying the same in solution as it is in the solid state. For instance, random-coil proteins like the chromogranin A proteins from the vesicles of the adrenal glands of animals would have to adopt some conformation if they were forced into a crystal state.

Lipscomb: Can they be crystallized?

Williams: Of course, they cannot be crystallized! But that is one of the problems: Professor Richards can talk only of crystallizable proteins. However, even some crystalline proteins (e.g. phospholipase A_2) do not conform to the accepted views drawn from the generalized studies of crystal structures of globular proteins. Thus phospholipase A_2 undergoes a continuous change in its structure from 0 to 95 °C (continuous on the n.m.r. time scale — i.e. it is altering its conformation at time intervals faster than 10^{-5} s). There is no single 'conformation' of the main chain. This may be of the greatest functional consequence.

Richards: What are you actually observing in the n.m.r. spectrum?

Williams: We see the positions of lines in the spectrum which are due to atomic environments. The lines are shifted by, for example, ring currents of neighbouring groups. The constancy of the ring-current shifts tells us whether the conformational ensemble is fixed. In other words, instead of using X-ray diffraction as a probe, we are using a tiny magnet (e.g. the benzene ring) to probe local regions. If the magnet is near a proton it will perturb its n.m.r. signal and so we can tell how close the magnet is to the proton.

Richards: How near is near?

Williams: Near means from 4 to around 10 Å; far means the distance in the totally denatured protein, in which distances between groups are great but limited in the sense that some proteins (e.g. phospholipase A_2) cannot denature totally because they contain S–S bridges. In phospholipase A_2 nearly all the ring-current shifts collapse at 95 °C — i.e. most of the n.m.r. spectra of the amino acids are such that the groups appear to be far from the aromatic rings. As the temperature is lowered to 0 °C the protein is continuously altering its structure — the ring currents are increasing — and even at 0 °C it has not reached a limiting conformation. I believe that this protein has a small core with floppy outside chains. There are then all degrees of behaviour between the maintenance of the solid-state single conformation and the grossly-averaging liquid state of random polymers when proteins are in solution. At one extreme cytochrome *c*, for example, shows slight internal motions which we would recognize probably at the level of the internal motions of a somewhat disordered crystal (very like a solid). (Compare the motion of ammonium when it starts to rotate in ammonium chloride; although the ammonium groups rotate freely, the structure is still a crystal.) In cytochrome c (Moore & Williams 1977) it is the phenylalanine rings which are restricted in their flipping rates at low temperatures. (Wüthrich & Sykes have made similar measurements on other proteins and M. Karplus has developed the theoretical side of this approach.) However, if one takes other proteins, the rates of flipping of the phenyl group are much faster. Varying extents and rates of motion are apparent in different proteins in solution.

Did I understand you to say that you could understand the properties of proteins in solution from the solid-state structure? I do not understand how that is possible in general.

Richards: No, we have evidence 'on both sides'; I was stating a problem rather than a solution. I do not understand why these flexible proteins do not form compact structures. The solvent-squeezing forces, regardless of how they arise or over what distance they are effective, appear to be large. There will be a tendency to form a compact structure. If phospholipase has these floppy 'wings' protruding from the compact core, what is the sequence that keeps them non-compact? In terms of the area change the forces tending to form a compact structure appear to be of the order of hundreds of kcal/mol. What could resist that force?

Williams: In general like charges resist packing — e.g. in histones, positively charged proteins. Again, consider phosvitin, a phosphorylated protein which has two negative charges for every third amino acid: that can hardly be expected to fold. Folding is controlled not just by the sum of the charges but by their location, i.e. the charge–charge interactions. In your calculations you considered only contact areas, not charges.

Richards: Yes; we took no account of charge. We are trying to dissect one aspect of the problem and to sort that out first.

Wüthrich: Slow exchange of the interior amide protons of the pancreatic trypsin inhibitor (Masson & Wüthrich 1973; Karplus *et al.* 1973; Wagner 1977) shows that the solution conformation of this globular protein is overall rather rigid. The n.m.r. data, however, provide evidence that the aromatic rings of Phe and Tyr rotate about the Cβ–Cγ bond (Wüthrich & Wagner 1975; Wagner *et al.* 1976). Since the aromatic rings are located in densely packed regions in the interior of the protein (Huber *et al.* 1970; Deisenhofer & Steigemann 1975), the question arose as to how the open space needed for the rotation of the rings becomes available. In the rigid X-ray structure rotation of the rings would be opposed by extremely-high energy barriers (Hetzel *et al.* 1976). In a theoretical model study starting from the energy-refined X-ray structure (Levitt 1974), we varied χ^2 for individual aromatic rings and then computed the conformational energies obtained in these modified structures after 100 cycles of energy minimization (Hetzel *et al.* 1976). These model studies gave evidence for reasonable values for the energies of activation for aromatic 'ring flips' and indicated that, in a flexible protein conformation, the perturbations of the surroundings by the flipping aromatic rings do not penetrate far into the protein; they are contained within spheres of radius 6–8 Å about the centre of the aromatic ring considered. We are now repeating these measurements but under high pressure to check whether we can get activation volumes for the ring flips in the protein.

Franks: Simple calculations can give us the magnitude of the mean-square fluctuations of volume and energy in the proteins. What you say is consistent with this; there are remarkably large fluctuations (calculated from the heat capacity and compressibility) in both energy and density (Cooper 1976).

Richards: One inconsistency which Professor Williams mentioned arises from G. Webber's measurements, for example, of the permeability of a protein (in a molecule, not the crystal) to oxygen and even bigger substances. Yet, in basic pancreatic trypsin inhibitor these protons in the main chain are exchangeable (on a time scale of months or years). If the protein fluctuations are as large as they appear to be to accommodate both ring motions and the permeability to oxygen, it is remarkable that some parts of that protein are so non-fluctuating that those protons do not exchange with water.

Wüthrich: We have studied a series of chemically modified forms of the basic pancreatic trypsin inhibitor as well as several homologous inhibitors. From comparison of over 200 resolved n.m.r. signals, we found that essentially-identical solution conformations prevail for all the different species. Therefore we decided that it would be particularly valuable to compare the temperature of denaturation, the N–H exchange rates and the rates of rotation for the aromatic rings. These experiments clearly showed that, although the stability with respect to thermal denaturation and the exchange rates of the amide protons were closely related, the aromatic ring flips appeared to be independent of the other two quantities. Typically, in a comparison with the native protein the denaturation temperature was lowered by 30 °C and the exchange rates of the amide protons were increased by three to four orders of magnitude, whereas the frequencies of the aromatic ring flips were essentially unaffected by the protein modification (G. Wagner & K. Wüthrich, unpublished results).

Lipscomb: How much fluctuation of the rotation of an aromatic would destroy the ring current which influences a neighbouring isoleucyl group?

Williams: We were measuring the flip time; a fast flip does not destroy the ring current. Complete averaging (fast uninhibited rotation) would reduce the ring current to a small value.

Lipscomb: In other words, if the ring-current effects disappear on a neighbouring isoleucyl group, do you suspect larger conformational changes than just rotation of the ring?

Williams: Yes. The rotations (flips) would have to be fast (no barriers) to have all orientations equal in occupancy. Such flippings could not occur even in the most simple molecules.

Roberts: I agree that the aromatic rings are 'flipping' between two orientations separated by 180°, rather than rotating and thus sampling all orientations. One *can* sometimes see rotational effects in the n.m.r. spectrum. For example, in the spectrum of dihydrofolate reductase we see the usual group of high-field methyl resonances, two of which we believe come from the same valyl residue. At low temperature these two methyl resonances have different ring-current shifts but, as the temperature is raised, they move together and become effectively superimposed. This occurs at temperatures at which the protein is clearly still in a folded state. This behaviour could be explained if there was rotation about the Cα–Cβ bond of the valyl residue at high temperature, so that the environments of the two methyl groups became averaged.

This is relevant to Professor Williams' point about phospholipase and the temperature-dependence of ring-current shifts. This temperature-dependence is seen, to a varying extent, in many proteins. It need not necessarily imply a major fluctuation in the structure with temperature — relatively local 'breathing' processes would be sufficient. Just as these allow 'flipping' of the aromatic rings, so they would allow rotational motion of side-chains bearing methyl groups. This averaging would lead to a reduction in the observed ring-current shift and hence to the observed temperature-dependence, without needing a major change in the structure (although this may happen in some cases).

Williams: In reduced cytochrome *c* some of the ring currents in the centre of the protein show practically no temperature-dependence. That means that the centre of cytochrome *c* is tight, up to about 90 °C. In lysozyme all the ring currents collapse steadily with temperature by small amounts. This behaviour suggests that the whole protein expands a little with temperature before it denatures at about 70 °C. In phospholipase A_2 all the ring-current shifts collapse continuously with increasing temperature to nothing at about 95 °C; different parts of the protein collapse with different temperature coefficients. My interpretation is that effectively only one portion of this protein is rigid, as Professor Wüthrich described for the protein which is rigid in the middle. Phospholipase A_2 is more floppy; and chromogranins are very floppy proteins. I believe that this has extreme functional importance and that the dynamics have been 'deliberately built' into these proteins, just as the static structures are.

References

CLARK, A. H., FRANKS, F., PEDLEY, M. & REID, D. S. (1977) Solute interactions in dilute solutions II. *J. Chem. Soc. Faraday Trans. I 73*, 290-305

COOPER, A. (1976) Thermodynamic fluctuations in protein molecules. *Proc. Natl. Acad. Sci. U.S.A. 73*, 2740-2741

DEISENHOFER, J. & STEIGEMANN, W. (1975) Crystallographic refinement of structure of bovine pancreatic trypsin inhibitor at 1.5 Å resolution. *Acta Cryst. B 31*, 238-250

FRANKS, F. (1975) *Water, a Comprehensive Treatise* (Franks, F., ed.), vol. 4, chap. 1, Plenum Press, New York

FRANKS, F., PEDLEY, M. & REID, D. S. (1976) Solute interactions in dilute solutions I. *J. Chem. Soc. Faraday Trans. I, 72*

HETZEL, R., WÜTHRICH, K., DEISENHOFER, J. & HUBER, R. (1976) Dynamics of aromatic amino-acid residues in globular conformation of basic pancreatic trypsin inhibitor. 2. Semiempirical energy calculations. *Biophys. Struct. Mech. 2*, 154-180

HUBER, R., KUKLA, D., RÜHLMANN, A., EPP, O. & FORMANEK, H. (1970) The basic trypsin inhibitor of bovine pancreas. I. Structure analysis and conformation of the polypeptide chain. *Naturwissenschaften 57*, 389-392

KARPLUS, S., SNYDER, G. H. & SYKES, B. D. (1973) A nuclear magnetic resonance study of bovine pancreatic trypsin inhibitor. Tyrosine titrations and backbone NH groups. *Biochemistry 12*, 1323-1329

LEVITT, M. (1974) Energy refinement of hen egg-white lysozyme. *J. Mol. Biol. 82*, 393-420

MASSON, A. & WÜTHRICH, K. (1973) Proton magnetic resonance investigation of the conformational properties of the basic pancreatic trypsin inhibitor. *FEBS (Fed. Eur. Biochem. Soc.) Lett. 31*, 114-118

MOORE, G. R. & WILLIAMS, R. J. P. (1977) Structural basis for the variation in redox potential of cytochromes. *FEBS (Fed. Eur. Biochem. Soc.) Lett. 79*, 229-232

PACKER, K. J. (1977) The dynamics of water in heterogeneous systems. *Philos. Trans. R. Soc. Lond. B 278*, 59-87

WAGNER, G. (1977) *Konformation und Dynamik von Protease-Inhibitoren: ¹H NMR Studien*, PhD Thesis (Nr. 5992), ETH Zürich

WAGNER, G., DE MARCO, A. & WÜTHRICH, K. (1976) Dynamics of aromatic amino-acid residues in globular conformation of basic trypsin inhibitor. 1. ¹H n.m.r. studies. *Biophys. Struct. Mech. 2*, 139-158

WÜTHRICH, K. & WAGNER, G. (1975) NMR investigations of the dynamics of the aromatic amino residues in the basic pancreatic trypsin inhibitor. *FEBS (Fed. Eur. Biochem. Soc.) Lett. 50*, 265-289

The chemical properties of out-of-equilibrium states of proteins and the role of these states in protein functioning

L. A. BLUMENFELD*

Institute of Chemical Physics, Academy of Sciences of the USSR, Moscow

Abstract   The out-of-equilibrium states of several iron-containing proteins (cytochromes *c* of different origin, haemoglobin, myoglobin, ferredoxin and other non-haem iron proteins, cytochrome *c* oxidase, horseradish peroxidase) were recorded after fast changes in the active centre (electron reduction of iron, ligand dissociation). Strained states result in which the active centre has already been changed and undergone vibrational relaxation but the main part of protein globule is in the 'old', now out-of-equilibrium, state. Protein structure and chemical properties in these states differ considerably from those in equilibrium states. As a rule, the rate constants of protein-specific chemical reactions increase in out-of-equilibrium states by 1–3 orders of magnitude in comparison with those in equilibrium states. Spectra and reactivity of these proteins change in the course of slow (up to 10^{-1} s) conformational relaxation, continuously approaching the equilibrium values. It seems that this conformational relaxation is essentially the elementary act of many enzymic reactions for which the rate of substrate–product transformation is determined by the rate of this conformational change.

Almost all protein chemical reactions, even if they affect only several groups in the active centre, are accompanied by considerable structural rearrangement of the whole protein macromolecule. The duration of these macromolecular rearrangements is, as a rule, much longer than that of the structural changes in compounds of low molecular weight and may be microseconds, milliseconds and even seconds. The realization of the transition between two quasi-equilibrium states after fast local perturbation (ligand binding to the active centre, redox change of the central metal atom, an acid group dissociation etc.) requires coordinated breaking and formation of many weak secondary bonds. Therefore a local chemical change in the active centre and

---

* Professor Blumenfeld was unfortunately unable to attend the symposium and, as his manuscript was not received until after the symposium, Professor Blow was asked at short notice to give a paper (see pp. 55-61). [Eds.]

fast vibrational relaxation accompanying this change ($\Delta\tau \approx 10^{-12} - 10^{-13}$ s) are realized before the structure of the main part of the protein globule has enough time to be changed. Between relaxed and unchanged regions of the protein structural strain appears, which vanishes only slowly in the course of protein conformational relaxation. Thus, after a chemical transformation of the active centre there appears a long-lived specific out-of-equilibrium state: the active centre is already changed, but the structure of the main volume of the protein macromolecule remains the same but is now out-of-equilibrium relative to the changed active centre. The appearance of these conformationally non-equilibrium states of individual molecules is a characteristic property of biopolymers — kinetically-frozen macromolecular constructions with a rigid memory on various levels of organization (Lifshitz 1968; Blumenfeld 1974, 1976a). These states determine the physical characteristics and chemical properties of a protein to a considerable extent. For many years we have systematically studied the conformationally out-of-equilibrium states of various metal-containing proteins and I shall summarize our main results in this paper.

In some experiments (Blumenfeld 1974; Blumenfeld et al. 1974a, b, c, 1975) the structurally non-equilibrium states were obtained as follows. The initial preparations were frozen water solutions of oxidized iron-containing proteins, the active centres of which may exist in oxidized as well as in reduced forms (cytochrome c of various origins, haemoglobin, myoglobin, iron–sulphur proteins, horseradish peroxidase, cytochrome oxidase). To obtain the rigid glass solutions at low temperatures one can add small quantities of ethylene glycol, glycerine, or saccharose. The active centres were reduced at 77 K with solvated electrons (obtained by radiolysis). The radiation dosage was chosen in such a way that the irradiation of already reduced proteins in the same conditions did not lead to any changes in spectroscopic and chemical properties of the proteins. In the frozen rigid matrix at 77 K the protein conformation cannot relax. Therefore, after electron reduction of the active centre there appears a stable (in rigid solution) out-of-equilibrium state of the type described above. The strain between the active centre and the apoprotein must influence the geometry and electronic characteristics of active centre. This strain is revealed in the optical and e.p.r. spectra of the active centre by comparison with the corresponding properties of the equilibrated reduced protein recorded in the same conditions (in frozen matrix). For example, in such conformationally-non-equilibrium reduced haemoglobin or myoglobin the haem group exists not in the high-spin but in the low-spin state and is characterized by a 'cytochrome-like' absorption spectrum. The width and intensity of e.p.r. signals from the iron–sulphur protein adrenodoxin are greatly changed in the out-of-equilibrium state. With increase of temperature the protein undergoes a relaxation to an equilibrium state identical to the usual one.

In the second type of experiments the conformationally out-of-equilibrium states of proteins were formed in water solutions at room temperature. The following methods were used: (1) the reduction of the active centre by microsecond pulses of fast electrons (Fel' *et al.* 1977; Davydov *et al.* 1977; Kuprin *et al.* 1977; Blumenfeld *et al.* 1977a); (2) flash photolysis of proteins of the carboxy-derivative type (Blumenfeld *et al.* 1977b); (3) the jump-like changes in pH values and in ionic strength (Greschner *et al.* 1976; Blumenfeld *et al.* 1976). In all these experiments the protein's electronic characteristics (absorption and fluorescence spectra) in the intermediate non-equilibrium states were recorded (beginning about 10 µs after the transformation of the active centre) and the kinetic parameters of individual relaxation stages were measured.

We have studied the following proteins: horse-heart cytochrome c and its derivatives; haemoglobin and myoglobin and their complexes with various low-molecular-weight ligands; ferredoxin; carboxyperoxidase. In all cases, electron reduction of active centre, ligand photodissociation, or pH jump leads to the rise of a state, the characteristics of which differ from those at equilibrium. The relaxation to the equilibrium state with gradual changes of spectroscopic properties takes a long time (up to hundreds of milliseconds). When we were able to measure the spectroscopic characteristics of the non-equilibrium states of reduced proteins in the initial relaxation stages, they did resemble the corresponding characteristics of states obtained by reduction of the active centre in a frozen matrix. The kinetic curves for the relaxation process can conveniently be presented as the sums of several exponents. The relaxation parameters are the values of first-order rate constants (k_i) and of the relative spectroscopic weights (A_i) of the individual relaxation stages. Table 1 gives the experimental values of k_i and A_i for some proteins.

The reactivity of proteins in conformationally out-of-equilibrium states was determined by measurement of rate constants of some specific reactions immediately after generation of these states and during their conformational relaxation. We have studied the following reactions: the oxidation of reduced cytochrome c, haemoglobin, myoglobin and ferredoxin by potassium ferricyanide and plastocyanin; the oxidation of reduced haemoglobin, myoglobin and ferredoxin by ferricytochrome c; the interaction of haemoglobin, myoglobin and their derivatives with molecular oxygen; the binding of carbon monoxide by peroxidase and by alkaline cytochrome c (see Table 2).

TABLE 1

The rate constants (k_i) and relative spectroscopic weights (A_i) of the relaxation stages for some metal-containing proteins

Protein	Concentration ($\mu mol/l$)	Temperature (°C)	pH	k_i/s^{-1}	$A_i(\%)$	Comments
Cytochrome c	8	20	7.15	2×10^4	30	
				2×10^3	40	
				20	30	
Haemoglobin	5	21	7.4	3.5×10^4	15	Conversion of low-spin form into high-spin one
				4×10^3	30 ⎫	Relaxation of
				4×10^2	30 ⎬	high-spin
				18	25 ⎭	form
Myoglobin	22	27	7.3	1.5×10^4	20	Conversion of low-spin form into high-spin one
				3×10^3	30 ⎫	Relaxation of
				6×10^2	10 ⎬	high-spin
				20	40 ⎭	form
Ferredoxin	85	25	7.2	4×10^3		
				6		

The main result of these kinetic studies is that in all cases the rate constants of the reactions of proteins in out-of-equilibrium states differ from those of the same proteins in equilibrium states. The rate constants change in the course of relaxation and the non-equilibrium proteins are usually more reactive. We have found only one exception: the rate of binding of carbon monoxide to horseradish peroxidase at room temperature measured immediately after the dissociation of CO by a powerful light flash is considerably lower during the first 100 ms than the CO-binding rate constant to the same protein in the equilibrium state.

The duration of the first rapid reaction stage is often close to that of the initial stage of structure relaxation after electron reduction of the active centre. In many cases, especially for reactions between proteins, the changes in reactivity continue long after the completion of the structural changes of the active centre recorded by spectra. When the duration of the reaction noticeably exceeds that of conformational relaxation, the rate constant of the last stage of the reaction coincides with that of protein in equilibrium state.

In all the protein reactions we studied two processes proceed simultaneously: the chemical reaction and the conformational relaxation of the

TABLE 2

The rate constants (k_i) and relative spectroscopic weights (A_i) of individual stages of the oxidation of some metal-containing proteins in out-of-equilibrium states measured immediately after reduction of active centres by solvated electrons

Protein	Oxidant	Temperature (°C)	pH	k_i/l $mol^{-1} s^{-1}$	A_i (%)	Corresponding rate constant for equilibrium state
20μM-Cytochrome c	50μM-Ferricyanide	25	7.1	10^9	15	
				4×10^7	85	5×10^7
30μM-Cytochrome c	50μM-Ferricyanide	25	10.5	10^8	70	
				1.5×10^7	30	2.8×10^5
80μM-Ferredoxin	80μM-Ferricytochrome c	22	7.1	2×10^8	50	
				2×10^7	20	8.1×10^4
				3×10^6	30	
0.1mM-Myoglobin	30μM-Ferricyanide	22	6.1	10^7	55	
				2.5×10^6	45	
0.1mM-Myoglobin	0.4mM-Ferricyanide	22	6.1	1.7×10^8	10	
				2×10^7	90	1.8×10^6
0.1mM-Myoglobin	0.8mM-Ferricyanide	22	6.1	10^8	20	
				2×10^7	80	
70μM-Myoglobin	40μM-Plastocyanin	23	7.2	2×10^7	10	
				1.5×10^6	20	
				1.5×10^5	70	
0.1mM-Haemoglobin	1mM-Ferricyanide	23	6.2	2×10^7	15	
				1.5×10^6	40	5×10^4
				4.5×10^4	45	

protein. Molecules of the same protein at different stages of relaxation are in fact different molecules with different electronic structures and different reactivity. This phenomenon leads to unexpected effects which as a rule cannot be observed for reactions of low-molecular-weight compounds. The effective rate constants of some chemical reactions of non-equilibrium proteins depend on the absolute rate of reaction which is determined by the concentration of reagents. This can be observed, for instance, for the reactions of haemoglobin and myoglobin with ferricyanide and oxygen: the effective rate constant increases with the concentration of low-molecular-weight reagent. With increases in reaction rate the greater part of the overall reaction has time to reach completion by the earliest stages of conformational relaxation, when the protein is most active. If the redox transformations are sufficiently fast, the protein molecules will be practically all the time in out-of-equilibrium states. It seems that these effects can explain the disagreement between the results of kinetic and thermodynamic studies of mitochondrial electron transport (see, for instance, Blumenfeld 1974) and the data concerning the increase of measured rate constants of electron transfer between components in the mitochondrial respiratory chain with the rate of respiration (Kupriyanov et al. 1976).

The reaction rates of proteins in out-of-equilibrium states often display unusual temperature dependences. The protein configuration and the rate of conformational relaxation are, as a rule, more sensitive to temperature changes than is the rate of the chemical reaction studied. Therefore, for instance, with decreasing temperature the conformational relaxation slows down, and the greater part of the reaction has enough time to reach completion during the first relaxation stages. Thus, the measured effective rate constant increases — in apparent variance with the usual Arrhenius dependence.

The appearance of non-equilibrium states of protein molecules with different kinetic characteristics was clearly demonstrated in the study of the reaction of haemoglobin with carbon monoxide under photolytic illumination (Blumenfeld et al. 1977b). Analysis of the transitions from stationary states in the light to the equilibrium states in the dark and vice versa at low partial pressures of carbon monoxide has led to the conclusion that there appear to be large quantities of protein molecules in out-of-equilibrium conformations with different rate constant values for their reaction with carbon monoxide. This calls into question the correctness of widely accepted schemes (Monod et al. 1965; Koshland et al. 1966) which postulate that the transitions of protein molecules between extreme conformations proceed instantly and, therefore, the intermediate conformational states do not appear.

These ideas led to the introduction of the relaxation concept of enzymic catalysis (Blumenfeld 1972, 1974, 1976a, b). This qualitative concept of the physical mechanism for the elementary act in enzymic catalysis is based on the postulate that the conformational changes of the substrate–enzyme complex accompanying the substrate binding to the active centre are of the nature of relaxation and include not only the breaking of old and formation of new secondary bonds in the protein macromolecule but also the chemical changes necessary to transform the substrate molecule into the molecule or molecules of product. This conformational relaxation is essentially the elementary act of enzymic catalysis, and the rate of substrate–product transformation is determined by the rate of this relaxation.

Certain quantitative theoretical approaches based on this concept have been suggested (Chismadzhev *et al.* 1976; Fain 1977). A similar approach was recently applied to muscle contraction (Gray & Gonda 1977) and to phosphorylation processes in membranes (Blumenfeld & Koltover 1972; Blumenfeld 1974, 1976*a, b*). These problems exceed, however, the limits of this report.

References

BLUMENFELD, L. A. (1972) [On elementary act in enzyme catalysis.] *Biofizika 17*, 954-959 [in Russian]

BLUMENFELD, L. A. (1974) *Problemy Biologicheskoj Fiziki*, Nauka, Moscow [English translation: *The Problems of Biological Physics*, Pergamon Press, 1978]

BLUMENFELD, L. A. (1976*a*) The physical aspects of enzyme functioning. *J. Theor. Biol. 56*, 269-284

BLUMENFELD, L. A. (1976*b*) [The physical aspects of intracellular energy transformation.] *Biofizika 21*, 946-957 [in Russian]

BLUMENFELD, L. A. & KOLTOVER, V. K. (1972) [The energy transformation and conformational transitions in mitochondrial membranes as relaxation processes.] *Mol. Biol. 6*, 161-166 [in Russian]

BLUMENFELD, L. A., BURBAEV, D. SH., VANIN, A. F., VILU, R. O., DAVYDOV, R. M. & MAGONOV, S. N. (1974*a*) [Out-of-equilibrium structures of enzyme metalloorganic centres.] *J. Strukt. Chim. 15*, 1030-1039 [in Russian]

BLUMENFELD, L. A., DAVYDOV, R. M., FEL', N. S., MAGONOV, S. N. & VILU, R. O. (1974*b*) Studies on the conformational changes of metalloproteins induced by electrons in water–ethylene glycol solutions at low temperatures. Cytochrome C. *FEBS (Fed. Eur. Biochem. Soc.) Lett. 45*, 256-258

BLUMENFELD, L. A., DAVYDOV, R. M., MAGONOV, S. N. & VILU, R. O. (1974*c*) Studies on the conformational changes of metalloproteins induced by electrons in water–ethylene glycol solutions at low temperatures. Hemoglobin. *FEBS (Fed. Eur. Biochem. Soc.) Lett. 49*, 246-248

BLUMENFELD, L. A., BURBAEV, D. SH., DAVYDOV, R.M., KUBRINA, L.N., VANIN, A.F. & VILU, R. O. (1975) Studies on the conformational changes of mtalloproteins induced by electrons in water–ethylene glycol solutions at low temperatures. Adrenodoxin. *Biochim. Biophys. Acta 379*, 512-516

BLUMENFELD, L. A., GRESCHNER, S., GENKIN, M. K., DAVYDOV, R. M. & ROLDUGINA, N. M. (1976) Kinetic study of conformational changes in ferricytochrome C induced by pH-change. *Stud. Biophys. 57*, 110

BLUMENFELD, L. A., DAVYDOV, R. M., KUPRIN, S. P. & STEPANOV, S. V. (1977*a*) [Chemical characteristics of conformationally non-equilibrium states of metal-containing proteins.] *Biofizika 22*, 977-994 [in Russian]

BLUMENFELD, L. A., ERMAKOV, YU. A. & PASECHNICK, V. I. (1977*b*) [Kinetics of haemoglobin – carbon monoxide reaction. 3. The existence of conformationally out-of-equilibrium Hb molecules during the processes of CO binding and dissociation.] *Biofizika 22*, 8-14 [in Russian]

CHISMADZHEV, YU. A., PASTUSHENKO, V. F. & BLUMENFELD, L. A. (1976) [On the dynamical theory of enzyme catalysis.] *Biofizika 21*, 208-213 [in Russian]

DAVYDOV, R. M., KUPRIN, S. P., FEL', N. S., POSTNIKOVA, G. B. & BLUMENFELD, L. A. (1977) [The study by the pulse radiolysis method of reactivity of conformationally out-of-equilibrium states of metalloproteins (myoglobin).] *Dokl. Akad. Nauk S.S.S.R. 235*, 950-952 [in Russian]

FAIN, V. M. (1976) On the theory of rate processes: the role of coherent mechanical vibrations. *J. Chem. Phys. 65*, 1854-1866

FEL', N. S., DOLIN, P. I., DAVYDOV, R. M., VANAG, V. K., KUPRIN, S. P., ROLDUGINA, N. M. & BLUMENFELD, L. A. (1977) [The study by the pulse radiolysis method of dynamical characteristics of conformationally out-of-equilibrium states of hemoglobin and its complexes.] *Elektrokhimiya 13*, 909-913 [in Russian]

GRAY, B. F. & GONDA, I. (1977) The sliding filament model of muscle contraction. 1. Quantum mechanical formalism. *J. Theor. Biol. 69*, 167-185

GRESCHNER, S., BLUMENFELD, L. A., GENKIN, M. V., DAVYDOV, R. M. & ROLDUGINA, N. M. (1976) Kinetics of conformational changes in acidic ferricytochrome *C* induced by salts. *Stud. Biophys. 57*, 109

KOSHLAND, D. E. JR., NEMETHY, G. & FILMER, D. (1966) Comparison of experimental binding data and theoretical models in protein containing subunits. *Biochemistry 5*, 365-385

KUPRIYANOV, V. V., POLOTCHIN, A. S. & LUSIKOV, V. N. (1976) [Analysis of stationary kinetics of uncoupled electron transfer through cytochrome chain in submitochondrial particles.] *Biokhimiya 41*, 1889-1897 [in Russian]

KUPRIN, S. P., DAVYDOV, R. M. FEL', N. S., NALBANDIAN, R.M. & BLUMENFELD, L. A. (1977) [The study by the pulse radiolysis method of reactivity of conformationally out-of-equilibrium states of metalloproteins (cytochrome *C*). *Dokl. Akad. Nauk S.S.S.R. 235*, 1193-1195 [in Russian]

LIFSHITZ, I. M. (1968) [Some aspects of statistical theory of biopolymers.] *Zh. Eksp. Teor. Fiz. 55*, 2408-2422 [in Russian]

MONOD, J., WYMAN, J. & CHANGEUX, J. P. (1965) On the nature of allosteric transitions: a plausible model. *J. Mol. Biol. 12*, 88-118

Flexibility and rigidity in protein crystals

D. M. BLOW

Biophysics Section, Blackett Laboratory, Imperial College, London

Abstract   There is an increasing body of crystallographic evidence for disorder in parts of the main polypeptide chain of certain proteins. The zymogens of pancreatic serine proteinases and tyrosyl-tRNA synthetase are used as examples. It is suggested that reactivity and specificity are favoured by an appropriate rigid conformation, and that disorder may often be involved in some kind of control function.

We all deeply regret the absence of Professor Blumenfeld from this symposium, and we feel his absence more acutely because only 24 hours ago we believed that he would be able to attend. I must apologize that my paper has been hurriedly prepared and that it can make no attempt to take the place of the paper which Professor Blumenfeld would have given*.

I shall discuss an aspect of protein structure about which there has appeared to be some disagreement between different groups of workers in the past, although I think this disagreement has been more a matter of emphasis than of real controversy. Protein crystallography gives clear information about the positions of atoms within a protein structure, but this information is averaged over a long period of time. Nuclear magnetic resonance gives information about the movement and flexibility of groups of proteins, but this information is difficult to correlate with precise positional data.

It has been suggested that everybody tends to emphasize what he can observe: crystallographers observe that definite structures exist and, therefore, they emphasize the functional role of these static structures; n.m.r. spectroscopists observe evidence for flexibility and, therefore, they emphasize the importance of movement.

---

* Professor Blumenfeld's manuscript arrived after the symposium (see p. 47). [Eds.]

There has been a germ of truth in this in the past but, as criteria of interpretation improve, there is some convergence of these two extremes. Thus the measurement of definite distances between nuclei with unpaired spins and surrounding ligands by n.m.r. spectroscopy gives structural information, and crystallographers are becoming able to interpret the local resolution or 'sharpness' of electron density maps as positive evidence of the degree of order. This evidence is most clear and definite where some kind of comparison can be made between an ordered and disordered state.

Protein crystallographers have always been aware of the possibility of disorder in crystal structures. Thus, in the first high-resolution protein map the NH_2-terminal residue of myoglobin was absent (Watson 1969), and similar absences near chain termini were found in ribonuclease-S (Wyckoff *et al.* 1970) and α-chymotrypsin (Birktoft & Blow 1972). Structures determined recently have shown that much larger amounts of disorder can exist within a molecule, involving the main polypeptide chain in regions remote from chain termini. This level of disorder is much greater than the flexibility calculated from a theoretical model of pancreatic trypsin inhibitor by McCammon *et al.* (1977), which demonstrates side-chain flexibility in a case where the main chain is relatively rigid.

I shall illustrate this by examples of crystallographic studies which involve the work in my laboratory in one way or another. But in doing so, I want first to acknowledge the important contribution made to the understanding of these states by Huber's group in Munich, in their work on the intact antibody molecule (Colman *et al.* 1976) and on trypsinogen (Fehlhammer *et al.* 1977).

ZYMOGEN AND ENZYME IN THE SERINE PROTEINASES

Following on the work of Hess and others (Oppenheimer *et al.* 1966), we were able to indicate the structural basis of the activation of the zymogens of the serine proteinases (Sigler *et al.* 1968). The cleavage which activates the zymogen releases a free α-amino group, which interacts, in the structure of the active enzyme, with Asp-194, adjacent to the catalytically-active Ser-195. We suggested that the activation of the zymogen was a parallel structural reorganization to that which occurs in active serine proteinases at high pH, resulting in a structural change with concomitant loss of catalytic activity. Further analysis of this structural change by Fersht & Requena (1971) showed that the equilibrium between active and inactive forms, in α-chymotrypsin, is such that even at the pH of highest activity only 85% of the enzyme is in active form, in solution.

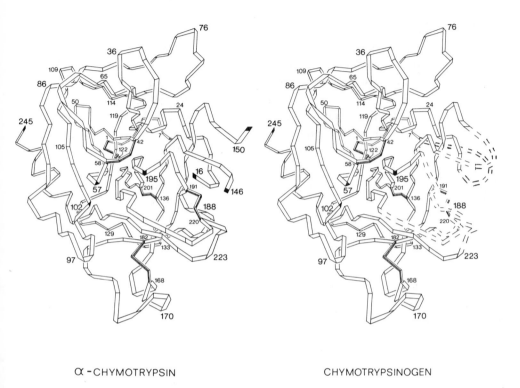

α - CHYMOTRYPSIN   CHYMOTRYPSINOGEN

FIG. 1. A structure for chymotrypsinogen, obtained by assuming disorder in the residues homologous to those found to be disordered in trypsinogen (Fehlhammer *et al.* 1977). Disordered
regions of the main chain are indicated by dashed lines. The structure of α-chymotrypsin (Birktoft
& Blow 1972), in which the only disordered segment of main chain is residues 7–13, is presented for
comparison.

It was assumed that this transition was a transition between two ordered
states of the enzyme. Freer *et al.* (1970) published the structure of bovine
chymotrypsinogen, showing that in the zymogen the substrate-binding pocket
did not exist. Although the catalytically-active residues of the charge-relay
system were all in place, one of the hydrogen bonds involved in binding the
carbonyl oxygen atom of the fissile peptide bond was unavailable.

In their recent analysis of the structure of bovine trypsinogen, Fehlhammer
et al. (1977) have shown clearly and unequivocally that a substantial proportion of the zymogen molecule is in a disordered state, and it seems likely that
this observation applies equally well to chymotrypsinogen. Fig. 1 shows an
illustration of the structure of chymotrypsinogen, obtained by assuming that
the disordered residues in chymotrypsinogen are homologous to those which

are found to be disordered in trypsinogen. A large surface of the molecule, stretching from the site of activation at the back of the molecule, all over the lower right region of its surface, including the peptide 147–148 which is removed from α-chymotrypsin, and two chains which construct the substrate binding pocket, are all in a state of disorder. P. Schwager (cited by Fehlhammer *et al.* 1977) has recently found that trypsinogen can be brought into the active conformation by complexation, either by complexing it with a trypsin inhibitor or by allowing it to bind the dipeptide Ile-Val.

These results favour the view that the active form of an enzyme, if it is to bind the substrate without unfavourable increase of entropy, must pre-exist in a shape which will bind the substrate favourably and without the need for excessive re-ordering of its peptide chains.

TYROSYL-tRNA SYNTHETASE

Tyrosyl-tRNA synthetase catalyses a series of reactions involving at least three substrates in an ordered reaction (see 1 and 2). There are important side-reactions which the enzyme catalyses, of which the most important is the hydrolysis of mischarged aminoacyl-tRNATyr compunds.

$$E + Tyr + ATP \rightarrow E:Tyr:AMP + PP_i \tag{1}$$
$$E:Tyr:AMP + tRNATyr \rightarrow E + Tyr-tRNATyr + AMP \tag{2}$$

The crystal structure of tyrosyl-tRNA synthetase from *B. stearothermophilus* is partially complete (Irwin *et al.* 1976). We find that the specificity of this enzyme is not achieved by a super-rigid, ideally engineered tyrosine-binding site. Binding experiments have shown that in the absence of tyrosine ATP binds with its ring in the tyrosine site, with a dissociation constant only 100 times that for tyrosine (C. Monteilhet & D. M. Blow, unpublished results). Each of the peptide chains of this dimeric molecule has a molecular weight of about 45 000 but the crystal electron-density map shows only 276 amino acids in a well-ordered state. We feared at first that the failure to interpret the remaining part of the molecule might be a result of the poor quality of the electron-density map at 2.7 Å resolution but the exceptionally high quality of the difference maps showing various types of ligand binding convinces us that the quality of the electron-density map is good. This suggests that part of the structure is highly ordered but the remainder of the molecule, which can only be detected as vague 'clouds' of electron density, is disordered.

There may be several conformational states for this enzyme, involving different ordering of different parts of its peptide chain, which may be involved in the expression of different types of catalytic activity.

CONCLUSION

In each case, it seems that a highly immobilized structure exists at the active site of an active form of the enzyme, providing a correctly formed surface which will give the necessary degree of reactivity and specificity. However, it seems that requirements of control, which demand that some active sites be non-functional or express a different activity in some conditions, can be implemented by specific transitions which involve a change in the degree of order of the enzyme.

References

BIRKTOFT, J. J. & BLOW, D. M. (1972) Structure of crystalline α-chymotrypsin. V. The atomic structure of tosyl-α-chymotrypsin at 2 Å resolution. *J. Mol. Biol.* 68, 187-240

COLMAN, P. M., DEISENHOFER, J., HUBER, R. & PALM, W. (1976) Structure of the human antibody molecule Kol (immunoglobulin G1): an electron density map at 5 Å resolution. *J. Mol. Biol.* 100, 257-282

FEHLHAMMER, H., BODE, W. & HUBER, R. (1977) Crystal structure of bovine trypsinogen at 1.8 Å resolution. II. Crystallographic refinement, refined crystal structure and comparison with bovine trypsin. *J. Mol. Biol.* 111, 415-438

FERSHT, A. R. & REQUENA, Y. (1971) Equilibrium and rate constants for the interconversion of two conformations of α-chymotrypsin. The existence of a catalytically inactive conformation at neutral pH. *J. Mol. Biol.* 60, 279-290

FREER, S. T., KRAUT, J., ROBERTUS, J. D., WRIGHT, H. T. & XUONG, NG. H. (1970) Chymotrypsinogen: 2.5 Å crystal structure, comparison with α-chymotrypsin, and implications for zymogen activation. *Biochemistry 9,* 1997-2009

IRWIN, M. J., NYBORG, J., REID, B. R. & BLOW, D. M. (1976) The crystal structure of tyrosyl-transfer RNA synthetase at 2.7 Å resolution. *J. Mol. Biol.* 105, 577-586

McCAMMON, J. A., GELIN, B. R. & KARPLUS, M. (1977) Dynamics of folded proteins. *Nature (Lond.)* 267, 585-590

OPPENHEIMER, H. L., LABOUESSE, B & HESS, G. P. (1966) Implication of an ionizing group in the control of conformation and activity of chymotrypsin. *J. Biol. Chem.* 241, 2720-2730

SIGLER, P. B., BLOW, D. M., MATTHEWS, B. W. & HENDERSON, R. (1968) The structure of crystalline α-chymotrypsin. II. A preliminary report including a hypothesis for the activation mechanism. *J. Mol. Biol.* 35, 143-164

WATSON, H. C. (1969) The stereochemistry of the protein myoglobin. *Progr. Stereochem. 4,* 299-333

WYCKOFF, H. W., TSERNOGLOU, D., HANSON, A. W., KNOX, J. R., LEE, B. & RICHARDS, F. M. (1970) The three-dimensional structure of ribonuclease-S. Interpretation of an electron density map at a nominal resolution of 2 Å. *J. Biol. Chem.* 245, 305-328

Discussion

Lipscomb: You mentioned that you find some of the larger structures with many ordered regions and a substantial amount of disordered regions. Another example is aspartate carbamoyltransferase, which has a highly ordered catalytic region and somewhat disordered regulatory regions. Crystallographers tend to generalize from a few instances but I suspect from this limited number of examples that the occurrence of both highly ordered and disordered regions may be encountered more often in larger proteins.

Phillips: Another example is phosphorylase *b*, in which the 15 residues at the NH_2-terminus of the chain cannot be seen in the crystal structure (Johnson *et al.* 1978) whereas in phosphorylase *a* that part can be seen (Fletterick *et al.* 1976).

Lipscomb: The same is true for the protein subunits in bushy stunt virus (S. C. Harrison, unpublished results, 1977).

Williams: N.m.r. spectroscopy sometimes gives information only about the converse: the mobile part is observable but not the fixed part. In soya-bean lipoxygenase (molecular weight 100 000) about half the protein appears to be random coil (very fast moving) but the rest cannot be observed (R. Egmond & R. J. P. Williams, unpublished results).

Richards: What is the present state of the work on the structure of immunoglobulin, with regard to size and disorder?

Lipscomb: Huber has had to piece together the structure from the fragments.

Blundell: In the Kol IgG immunoglobulin that Huber *et al.* (1976) studied the *Fc* fragment cannot be seen in the electron-density map and is disordered, but the *Fab* parts are ordered. On the other hand, Silverton *et al.* (1977) find that the *Fc* fragment is ordered. Of course, there is a deletion in the hinge region of this immunoglobulin which may make it less flexible.

Richards: If large well-ordered areas were connected together by relatively short and flexible tethers, the resulting structure would have different properties from one with large regions of chain which are literally random coil.

Blow: It is not necessarily random coil, because each piece of disordered chain is short — only half a dozen peptide bonds between two ordered lengths of chain.

Richards: Those people who worry about the folding of proteins and whether one can predict how they fold in the three-dimensional sense have been under the impression that they were trying to define the final stable structures. According to current dogma, these structures are presumably controlled by sequence. If a major part of the problem is the definition of sequences that are unable to take up such structures, then we have an added degree of complexity which I must confess I had not previously considered.

Williams: Another difficulty besides non-folded proteins is exemplified by a ferredoxin, isolated by Xavier, which contains several identical subunits. When the subunits come together, they form two different oligomers, each with completely different properties around the iron–sulphur bonds. The redox potentials are altered by about 0.3 V. The two oligomers are both dimers or trimers. Apparently the one subunit may fold in two ways in the oligomers.

Blow: Another surprising feature is that a small perturbation can cause profound folding transitions, so that even if we could predict one structure, we would still be nowhere near defining its stability.

Richards: Undoubtedly that is a difficult problem but we may find considerable conservation of structure, even between apparently very different conformers.

Vallee: Is the lack (or amount) of discernible structure necessarily associated with the size of the protein that is being studied?

Lipscomb: Not necessarily, but it is often correlated with structure. In these examples it is.

Vallee: Is it also seen in smaller proteins?

Lipscomb: A disordered loop is often seen, even in small proteins.

References

FLETTERICK, R. J., SYGUSCH, J., SEMPLE, M. & MADSEN, N. B. (1976) Structure of glycogen phosphorylase *a* at 3.0 Å resolution and its ligand binding sites at 6 Å. *J. Biol. Chem. 251*, 6142-6146

HUBER, R., DEISENHOFER, J., COLMAN, P. M., MATSUSHIMA, M. & PALM, W. (1976) Crystallographic structure studies of an IgG molecule and an *Fc* fragment. *Nature (Lond.) 264*, 415-419

JOHNSON, L. N., WEVER, I. T., WILD, D. L., WILSON, K. S., & YEATES, D. G. R. (1978) in *Proceedings of 11th FEBS (Fed. Eur. Biochem. Soc.) Symposium* (Copenhagen 1977), Pergamon, Oxford, *42*, 185-194

SILVERTON, E. W., NAVIA, M. A. & DAVIES, D. R. (1977) Three dimensional structure of an intact human immunoglobulin. *Proc. Natl. Acad. Sci. U.S.A. 74*, 5140-5144

Coenzyme-induced conformational changes and substrate binding in liver alcohol dehydrogenase

CARL-IVAR BRÄNDÉN and HANS EKLUND

Department of Chemistry, Swedish University of Agricultural Sciences, Uppsala

Abstract   The apoenzyme and holoenzyme structures of liver alcohol dehydrogenase have been determined by X-ray methods to obtain details about coenzyme binding, substrate specificity and the catalytic mechanism. Coenzyme binding induces a conformational change of the protein which partly shields the active site from the solution. The reduced coenzyme binds in an open conformation similar to that of NAD bound to malate dehydrogenase. A hydrogen bond between Thr-178 and the carboxamide group of the coenzyme is essential for proper positioning of the nicotinamide in the active site. Coenzyme analogues in which the carboxamide group is absent or substituted with iodine bind in a different conformation and do not induce the structural change of the protein. Binding of substrate molecules has been studied in crystals obtained from an equilibrium mixture of enzyme, coenzyme and *p*-bromobenzyl alcohol. The oxygen atom of this substrate as well as that of the inhibitor molecules trifluoroethanol and dimethyl sulphoxide bind directly to the catalytic zinc atom. The substrate-binding region is a deep hydrophobic pocket at the bottom of which the zinc atom mediates electrophilic catalysis of alcohol oxidation.

Studies of alcohol dehydrogenases (EC 1.1.1.1) have been important in the development of some general concepts in enzymology. Yeast alcohol dehydrogenase was used by Westheimer and his colleagues to demonstrate direct transfer of hydrogen between the substrate and the pyridine nucleotide coenzyme and that this transfer is stereospecific with respect to the nicotinamide ring. Subsequent studies have shown (Popják 1970) that all pyridine-nucleotide-dependent dehydrogenases can be divided into two groups, having either A- or B-stereospecificity, depending on which side of the nicotinamide ring hydride transfer occurs.

Kinetic studies on liver alcohol dehydrogenase provided the experimental basis for Theorell & Chance (1951) to formulate the theories for an ordered mechanism. Binding of oxidized coenzyme precedes productive alcohol binding and dissociation of reduced coenzyme is the last and rate-limiting step of the reaction sequence. It has since been shown that many enzymes use this type of mechanism.

Alcohol dehydrogenase from liver was used by Prelog (1964) to develop the diamond-lattice concept of substrate specificity. He showed that by studying the turnover rates of saturated cyclic substrates of known or predictable conformations he could define the space available for productive substrate binding in terms of allowed or forbidden diamond-lattice points. These investigations have later been extended by Dutler (1977) and further explored for use in synthetic organic chemistry (Jones & Beck 1976).

Transient kinetic studies of liver alcohol dehydrogenase in recent years have raised the question of whether this enzyme exhibits half-site reactivity. The problem is mainly whether the two subunits of the dimeric molecule operate independently of each other or whether the catalytic activity of one subunit is influenced by the state of the other subunit.

We started X-ray studies of liver alcohol dehydrogenase hoping that the structure determinations of this enzyme might provide some understanding of the molecular basis of the above-mentioned concepts. In addition there were several unsettled questions regarding the specific catalytic mechanism of this enzyme. We wanted to know the role of the catalytically-important zinc atom and why it is needed in this particular dehydrogenase and not in others. Furthermore, in spite of the fact that the enzyme has a broad substrate specificity towards both primary and secondary alcohols, it does not oxidize several important cell metabolites that contain alcohol groups, such as lactate, malate and sugar molecules. There is thus some stringent discrimination in spite of the broad substrate specificity. Finally, hydride transfer from primary alcohols like ethanol is stereospecific in spite of the fact that secondary alcohols such as cyclohexanol are good substrates (Dalziel 1975). The stereospecificity of ethanol oxidation thus cannot be due to steric hindrance for binding the ethanol molecule in the wrong orientation.

We have determined the structure of the apoenzyme at high resolution (Eklund et al. 1976) and studied the binding of inhibitor molecules. Some of these are coenzyme analogues which bind to the enzyme but do not induce the structural transition to the holoenzyme conformation (Abdallah et al. 1975; Samama et al. 1977).

In collaboration with J.-F. Biellmann and his colleagues in Strassbourg we have determined which groups on the coenzyme are necessary to induce the conformational change. We have also determined the structure of the holoenzyme at low resolution (H. Eklund & C.-I. Brändén, unpublished work) and studied the binding of substrates and substrate competivive inhibitors (B.

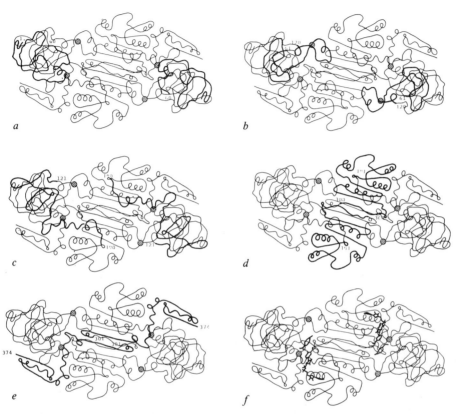

FIG. 1. Schematic diagram of the positions of different regions of the polypeptide chain: the shaded circles are zinc atoms; thin lines represent the whole molecule; thick lines represent in (a) residues 1–60; (b) residues 61–120; (c) residues 121–190; (d) residues 191–300; (e) residues 301–373; (f) bound coenzyme and substrate.

Plapp *et al.*, unpublished work) to this conformation. We shall summarize here these results and discuss the implications of the structural studies for the problems outlined above.

APOENZYME STRUCTURE

In the absence of coenzyme we always obtain orthorhombic crystals of the apoenzyme. Their structure has been determined at a resolution of 2.4 Å (Eklund *et al.* 1976) using the known amino acid sequence (Jörnvall 1970). Fig. 1 illustrates the positions of different regions of the two crystallographically-identical polypeptide chains in the dimeric molecule. Each subunit is divided into two domains, one of which binds the coenzyme, NAD. This

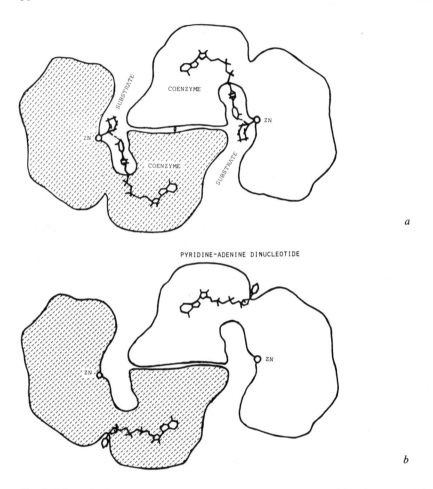

FIG. 2. Schematic diagram of a section through the middle of the alcohol dehydrogenase molecule. Projected onto this section are (*a*) bound coenzyme and substrate, and (*b*) bound pyridine-adenine dinucleotide.

domain comprises residues 176–318 and has a structure which is homologous to corresponding NAD-binding domains in other dehydrogenases (Rossmann *et al.* 1975). The second domain provides the residues which bind the two zinc atoms of the subunit. The two subunits are connected in the middle of the molecule by interactions between the coenzyme-binding domains as illustrated in Figs. 1*d* and *e*. The catalytic zinc atom is in the region between the two domains at the bottom of a deep pocket which is lined mainly by hydrophobic residues. This is schematically illustrated in Figs. 2 and 3 which are based on sections through a computer-generated space-filling represen-

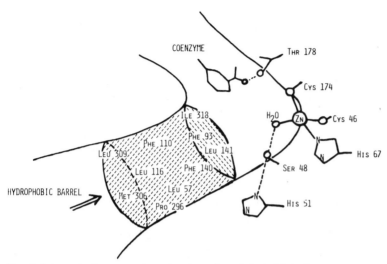

FIG. 3. Schematic diagram of the active site region of alcohol dehydrogenase.

tation of the molecule. (These calculations were made in collaboration with B. K. Lee.) The zinc atom is liganded to three protein groups: Cys-46, Cys-174 and His-67. In addition a water molecule which projects out into the pocket is also bound to the zinc atom. This water molecule is connected to the surface of the protein molecule through a hydrogen-bond system involving Ser-48 and His-51. Inhibitor molecules such as imidazole and 1,10-phenanthroline are bound in the deep pocket liganded to the zinc atom thereby displacing the water molecule (Boiwe & Brändén 1977). The zinc atom is pentacoordinate in the 1,10-phenanthroline complex.

HOLOENZYME STRUCTURE

With the coenzyme the holoenzyme gives triclinic crystals. The structure of the complex between enzyme, reduced coenzyme and dimethyl sulphoxide has recently been determined at a resolution of 4.5 Å (H. Eklund & C.-I. Brändén, unpublished work). In these crystals the subunits are not crystallographically identical but occupy independent positions in the asymmetric unit. Phase angles for this map were obtained by isomorphous replacement independent of the apoenzyme structure. The interpretation of this map was, however, greatly facilitated by projecting the known apoenzyme structure onto the holoenzyme map. By superimposing the known apoenzyme atomic model, a 4.5 Å apoenzyme map, and the 4.5 Å holoenzyme map section by section, we obtained a detailed interpretation in terms of the holoenzyme structure.

The structures of the two independent subunits of the holoenzyme molecule are similar. We can observe some systematic differences, especially in the active site regions which are of the order of 1–2 Å units. We cannot, however, determine at this resolution whether these differences are significant or artifacts of the low resolution.

There are significant conformational differences between the apoenzyme and holoenzyme subunit structures although the overall fold of the polypeptide chains is similar. The conformational differences can be described as a combination of a twist of the whole catalytic domain with respect to the coenzyme-binding domain and small local structural changes of the path of the polypeptide chain. One effect of the twist is to narrow the gap between the domains. The largest local conformational difference is also seen in the region between the domains and involves residues 51–60.

Residues 46–55 are arranged in a helix in the apoenzyme, whereas residues 56–60 have no regular secondary structure and comprise a loop between this helix and a strand in the pleated-sheet region in the catalytic domain. In the holoenzyme structure this helix is shorter and has a somewhat different direction. The last residues of the helix are unwrapped and are part of the loop region. Furthermore, the tip of this loop has moved closer to the coenzyme-binding domain, filling up a gap between the domains. The magnitude of the movement is large, about 6 Å for the tip of the loop. By this movement the loop is positioned outside the nicotinamide and nicotinamide-ribose part of the coenzyme. The combined effect of coenzyme binding, twist of the domains and loop movement is completely to shield off access to the active site from this direction (perpendicular to the plane of the paper in Fig. 2). We have tried to illustrate these movements schematically in Fig. 4.

A comparable significant loop movement on binding of the coenzyme has been observed in lactate dehydrogenase (EC 1.1.1.27) where a loop in the coenzyme-binding domain moves closer to the nicotinamide end of the coenzyme (White et al. 1976). A comparison between the loop movements in these dehydrogenases shows that although the loops are parts of non-homologous regions in these enzymes, the positions of these loops relative to the coenzyme position after the loop movements seem to be similar in space. Thus we may here have an example of functional convergence where structurally different loops have moved into similar positions relative to the active site producing similar effects, namely the shielding of the active site from the solution.

The movements of the tryptophyl residues are of interest in connection with protein-fluorescence studies. Each subunit has only two tryptophyl residues – Trp-15 and Trp-314. Trp-314, which is completely buried in a hydrophobic

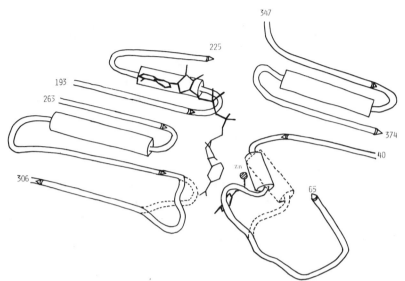

FIG. 4. Schematic diagram of some of the conformational differences between apoenzyme and holoenzyme in the region between the two domains of the subunit: ———, holoenzyme; ----, apoenzyme.

environment, does not move within the resolution we observe. Trp-15, which is more accessible to the solution, changes position owing to the twist of the catalytic domain and also seems to have changed orientation and local environment.

Compared to the apoenzyme structure no new residues have been brought into the active site. Furthermore, the positions of the protein groups which have been suggested (Brändén *et al.* 1975) as important for catalysis and substrate binding have changed only marginally with the possible exception of His-51.

COENZYME BINDING

Electron densities corresponding to bound coenzyme are found in each of the subunits. These have roughly the same heights showing that the coenzyme molecules bind to the same extent in both subunits. Rotation of one of these two continuous chains of density around the local two-fold axis of the molecule shows that all the major maxima within the densities superimpose to within 0.5 Å except those corresponding to the nicotinamide groups which differ by 1.5 Å. This difference may or may not be significant at the present resolution.

A 4.5 Å resolution map does not allow a unique determination of the set of dihedral angles that describe the conformation of the coenzyme molecule. However, the positions of all parts of the dinucleotide must agree with corresponding maxima in the continuous chain of electron density. We found that the conformation observed for NAD+ (Banaszak & Webb 1975) bound to malate dehydrogenase (EC 1.1.1.37) could be fitted to our density with only minor adjustments. We have previously shown that ADP-ribose binds to the apoenzyme without inducing any observable conformational change (Abdallah *et al.* 1975). The mode of binding of ADP-ribose in the orthorhombic crystals and its interactions with the protein (Eklund *et al.* 1976) are similar to those of the corresponding part of the coenzyme observed in the triclinic holoenzyme crystals. The nicotinamide ring is positioned in a crevice between the two domains of the subunit rather deep inside the protein. One side of the ring faces a hydrophobic region of the coenzyme-binding domain and the other side faces the catalytic domain close to the zinc atom of the active site.

Since the nicotinamide group is essential for the coenzyme-induced conformational change we looked for possible specific interactions between the protein and the nicotinamide. The only polar side-chain in this region is the hydroxy group of Thr-178 which might form a hydrogen bond to the carboxamide group of the nicotinamide ring. To deduce the possible importance of this group we studied the binding of coenzyme analogues in which the carboxamide group has been either removed or substituted with an iodine atom (Samama *et al.* 1977). Neither analogue induces the conformational transition of the protein. Both analogues bind in the same way, which is completely different from that found for coenzyme binding in the triclinic crystals. The nicotinamide groups of the coenzyme and the analogues are in completely different positions, about 1.6 nm (16 Å) from each other. The adenine groups bind in the same positions. The mode of binding of these analogues is schematically illustrated in Fig. 2*b*.

In conclusion, the coenzyme molecule is positioned in an open conformation in a crevice of the NAD-binding domain by several contacts which extend over the whole coenzyme molecule. These contacts are more frequent in the AMP-part which is thus more firmly anchored to the enzyme than the nicotinamide mononucleotide part. A hydrogen bond between the protein and the carboxamide group of the coenzyme is essential for proper positioning of the nicotinamide ring in the active site and the accompanying conformational transition from apoenzyme to holoenzyme structure. In the absence of the carboxamide group the dinucleotide binds in a different way and no conformational change is observed.

SUBSTRATE BINDING

One can calculate from kinetic and association constants that, after equilibration with high concentrations of NAD+ and benzyl alcohol, more than 90% of the enzyme will be in the form of the ternary complex with these two substrates. B. P. Plapp has crystallized such an equilibrium mixture and, in collaboration with us (B. P. Plapp, H. Eklund & C.-I. Brändén, unpublished work), has determined the mode of binding of the substrate molecule in one such complex. We chose p-bromobenzyl alcohol as the substrate since the presence of a heavy atom in the substrate molecule would facilitate the interpretation of the low-resolution difference Fourier maps.

We obtained triclinic crystals of good quality that were isomorphous to the enzyme–NADH–dimethyl sulphoxide crystals. We do not know what oxidation states of coenzyme and substrate predominate in these crystals, but they are currently being investigated by E. Zeppezauer in collaboration with G. Rossi at Parma using his technique for kinetic measurements in single crystals. We do know, however, from our difference Fourier maps, that both coenzyme and the substrate molecules labelled with the heavy atom are bound to the enzyme. We have also in collaboration with Plapp studied triclinic crystals of binary complexes with NAD+ or NADH and ternary complexes from solutions of NAD+ and the inhibitor trifluoroethanol. Although studies at higher resolution are needed to confirm our conclusions, we feel confident about the following results from our interpretations of the difference Fourier maps between these complexes.

Each subunit binds one substrate or inhibitor molecule with roughly the same occupancy. The oxygen atom of the substrate molecule or trifluoroethanol is directly liganded to the active-site zinc atom in almost the same position as the zinc-bound water molecule in the apoenzyme structure. The oxygen atom of dimethyl sulphoxide, which might be considered as an aldehyde analogue, is also directly bound to zinc but in a different position — almost the same as the nitrogen atom of bound imidazole to the apoenzyme structure (Boiwe & Brändén 1977). The substrate carbon atom from which the hydrogen atom is transferred to the coenzyme is 3.5 Å from C4 in the models we have fitted to the densities. Almost the same position is found for both trifluoroethanol and the substrate. The bromo-substituted benzene ring of the substrate is bound in the hydrophobic pocket that was predicted to bind substrate molecules from the apoenzyme structure (Eklund et al. 1976). Fig. 5 shows a schematic diagram of zinc, substrate and coenzyme molecule.

When the oxygen, carbon and bromine atoms of the substrate are positioned in this way, which provides the best fit to the difference Fourier

NICOTINAMIDE

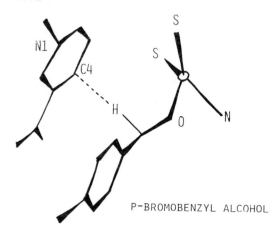

FIG. 5. Schematic diagram of zinc, substrate and the nicotinamide part of the coenzyme.

maps, the correct substrate stereochemistry for hydride transfer (Popják 1970) is automatically obtained. The C–H bond direction for the hydrogen atom to be transferred points toward C4 of the nicotinamide ring with an angle N1–C4–C(substrate) of 100°.

Having thus defined the substrate-binding position we can examine the properties of the substrate site and the effect of the coenzyme-induced conformational change on the availability of this site for substrates in solution. In the apoenzyme structure the catalytic zinc atom is accessible to molecules in solution from two different directions. Coenzyme binding and the conformational change closes off access to zinc from one of these directions. In the holoenzyme structure the catalytic site is accessible only through a deep and comparatively narrow pocket (Figs. 2 and 3). The zinc atom sits at the bottom of this pocket where the following polar groups are present: Thr-178, the carboxamide group of the coenzyme, Ser-48 and the zinc ligands. The pocket is lined with non-polar residues from both subunits which form a hydrophobic barrel between the active site and the surface of the molecule (Fig. 3). The size of the pocket can be visualized from model-building studies which show that the pocket is sufficiently deep to accommodate a steroid molecule with the 3-hydroxy group of the A-ring liganded to zinc and the terminal 26-carboxy group forming a salt bridge to an arginyl residue at the rim of the pocket. This molecule almost completely fills up the space available inside the pocket, with little steric interference with the protein as though the pocket were designed to accommodate steroid-like molecules.

For a small substrate molecule like ethanol to reach the active site from the solution, it must thus travel into the pocket, through the hydrophobic barrel to the bottom where it finds the zinc atom to which it is liganded and positioned properly for hydride transfer.

CONCLUSIONS

The ordered mechanism implies that the productive substrate binding site is formed after coenzyme binding. We have seen that coenzyme binding induces a conformational change. We know from solution studies that the affinity for small substrate molecules increases drastically after coenzyme binding. It remains to deduce why the conformational change increases this affinity. It turns out that there is no easily obtainable answer in structural terms. It is a question neither of providing space for the substrate, since space is available in the apoenzyme structure for the observed mode of substrate and inhibitor binding, nor of changing the orientation of some protein residues to provide specific interactions with the substrate. The substrate specificity is broad and apart from the zinc–oxygen bond there is no specific interaction between enzyme and substrate but rather several unspecific non-polar interactions. It seems to us that one of the main possibilities left is a subtle change in the electronic properties of the zinc atom. This change would increase the affinity of the zinc atom for oxygen ligands. Another possibility is that the protein's vibrational entropy contribution to substrate binding is more favourable in the holoenzyme structure than in the apoenzyme structure. Sturtevant (1977) has recently pointed out the importance of this term to the free energy of ligand binding to proteins.

The nature of the substrate-binding pocket provides a rational explanation for the broad substrate specificity of different aliphatic and aromatic alcohols and for the discrimination of alcohols containing charged groups. The hydrophobic pocket can accommodate a wide variety of non-polar side-chains by non-specific hydrophobic interactions. The only specific recognition site is the zinc atom which binds the alcohol group. Thus alcohol dehydrogenases with a broad substrate specificity need a metal atom to recognize and bind the substrate molecules. This is in contrast to other dehydrogenases such as lactate dehydrogenase which recognize both the alcohol group and adjacent charged groups. Such substrates are prevented from binding to alcohol dehydrogenase by the hydrophobic barrel in the substrate-binding pocket.

The stereospecificity of hydride transfer (Popják 1970) can, at least partly, be rationalized in structural terms. The A-side of the nicotinamide ring faces the substrate. The B-side is not accessible to the substrate since it is close to a

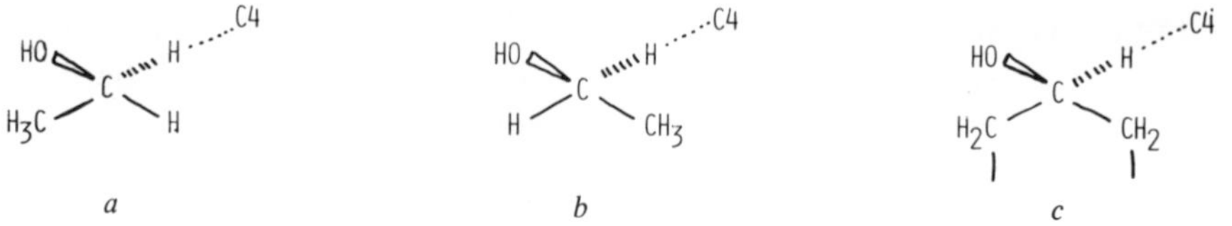

FIG. 6. Schematic diagram illustrating binding of primary and secondary alcohols: (a) ethanol with observed stereospecificity of hydride transfer; (b) ethanol with incorrect stereospecificity; (c) cyclohexanol.

hydrophobic region of the protein. The hydride transfer of the coenzyme must thus be stereospecific. The hand of this stereospecificity is determined by the position of the carboxamide group. If the nicotinamide ring were turned 180° around the glycosidic bond so that the B-side faces the substrate, the carboxamide group would come too close to Cys-174 and Gly-175. The binding of the nicotinamide in this conformation is thus prevented by steric interference with protein groups in the active site.

The substrate stereospecificity depends on more subtle factors. The problem has been recognized by kinetic studies and is schematically illustrated in Fig. 6. Why does bound ethanol have the CH_3 group in only one of the two possible orientations when secondary alcohols like cyclohexanol have CH_2 groups in both positions? Examination of the detailed apoenzyme structure using the positions observed for coenzyme and substrate binding in the holoenzyme shows that both possible CH_3 positions are sterically available for binding. However, the methyl group of the substrate is closer to the nicotinamide in the orientation that gives the correct stereochemistry for hydride transfer. Furthermore, the position for the methyl group in the other orientation seems to be more crowded. These factors might provide the energy needed to discriminate between the two possible orientations of primary alcohols. Secondary alcohols such as cyclohexanol can bind with CH_2 groups in both positions because hydrophobic interactions between the remaining CH_2 groups of the ring and atoms in the hydrophobic barrel provide further attraction energy for binding.

A comparison between the hands of nicotinamide and substrate in different dehydrogenases indicate that these hands may be related (M. G. Rossmann, P. Argos & C.-I. Brändén, unpublished work). The alcohol group, the hydrogen to be transferred and the direction of a substrate-binding position on the protein were used to define the hand of the bound substrate. Lactate and alcohol dehydrogenases which have A-specific hydride transfer to the coenzyme both have the same hand of bound substrate. In contrast, glyceraldehyde-3-phosphate dehydrogenase where hydride transfer to the nicotinamide is B-specific has an opposite hand of the substrate. In alcohol dehydrogenase the hydrophobic interaction between the methyl group of bound ethanol and the carboxamide carbon atom of the coenzyme could be partly responsible for this relation of stereospecificity between coenzyme and substrate.

We have assumed that all substrate molecules bind with the oxygen atom in the same place and with the same direction of the C–H bond of the transferred hydrogen. Using this assumption Prelog (1964) showed that the position of the carbon atoms of alicyclic substrates bound to the active site can be defined in terms of a diamond lattice. Dutler (1977) has recently made an extensive and careful kinetic study of methyl-substituted cyclohexanol derivatives. He has also, in collaboration with us, correlated his direct map of the space available for substrate binding with the X-ray structure of the substrate-binding pocket. The qualitative correlation is excellent (Dutler 1977).

The basic aspects of the catalytic mechanism of one subunit that we suggested (Brändén et al. 1975) on the basis of the apoenzyme structure have been confirmed by the recent studies of the holoenzyme structure. No additional protein group is brought into the active site by the conformational transition. Additional structural features of the mechanism, such as differences in the detailed geometries of the two central ternary complexes or the coordination of a water molecule to zinc in addition to the substrate oxygen atom, can, we hope, be obtained by the high-resolution studies of suitable holoenzyme complexes now in progress. It is also apparent that such studies are needed to determine whether the structural differences between the two subunits of the holoenzyme we have observed are artifacts of the low resolution or whether there is significant structural basis for the two subunits to operate with unequal efficiency.

ACKNOWLEDGEMENT

This work has been supported by the Swedish National Research Council (grant no. 2767).

References

ABDALLAH, M. A., BIELLMANN, J.-F., NORDSTRÖM, B. & BRÄNDÉN, C.-I. (1975) The conformation of adenosine diphosphoribose and 8-bromoadenosine diphosphoribose when bound to liver alcohol dehydrogenase. Eur. J. Biochem. 50, 475-481

BANASZAK, L. J. & WEBB, L. E. (1975) Nicotinamide adenine dinucleotide and the active site of cytoplasmic malate dehydrogenase, in Structure and Conformation of Nucleic Acids and Protein–Nucleic Acid Interactions (Sundaralingam, M. & Rao, S. T., eds.), pp. 375-386, University Park Press, Baltimore

BOIWE, T. & BRÄNDÉN, C.-I. (1977) X-ray investigation of the binding of 1,10-phenanthroline and imidazole to horse liver alcohol dehydrogenase. *Eur. J. Biochem.* 77, 173-179

BRÄNDÉN, C.-I., JÖRNVALL, H., EKLUND, H. & FURUGREN, B. (1975) Alcohol dehydrogenases, in *The Enzymes*, 3rd edn. (Boyer, P. D., ed.), vol. 11A, pp. 103-190, Academic Press, New York

DALZIEL, K. (1975) Kinetics and mechanism of nicotinamide-nucleotide-linked dehydrogenases, in *The Enzymes*, 3rd edn. (Boyer, P. D., ed.), vol. 11A, pp. 1-60, Academic Press, New York

DUTLER, H. (1977) Probing the active site of liver alcohol dehydrogenase. *Biochem. Soc. Trans.* 5, 617-620

EKLUND, H., NORDSTRÖM, B., ZEPPEZAUER, E., SÖDERLUNG, G., OHLSSON, I., BOIWE, T., SÖDER-BERG, B.-O., TAPIA, P., BRÄNDÉN, C.-I. & ÅKESON, Å. (1976) Three-dimensional structure of horse liver alcohol dehydrogenase at 2.4 Å resolution. *J. Mol. Biol.* 102, 27-59

JONES, J. B. & BECK, J. F. (1976) Asymmetric synthesis and resolutions using enzymes, in *Applications of Biochemical Systems in Organic Chemistry*, part 1 (Jones, J. B., Sih, C. J. & Perlman, D., eds.), (*Techniques of Chemistry Vol. X*), pp. 236-401, Wiley, New York

JÖRNVALL, H. (1970) Horse liver alcohol dehydrogenase. The primary structure of the protein chain of the ethanol-active isoenzyme. *Eur. J. Biochem.* 16, 25-40

POPJÁK, G. (1970) Stereospecificity of enzymic reactions, in *The Enzymes*, 3rd edn. (Boyer, P. D., ed.), vol. 2, pp. 115-215, Academic Press, New York

PRELOG, V. (1964) Specification of the stereospecificity of some oxido-reductases by diamond lattice sections. *Pure Appl. Chem.* 9, 119-130

ROSSMANN, M. G., LILJAS, A., BRÄNDÉN, C.-I. & BANASZAK, J. L. (1975) Evolutionary and structural relationships among dehydrogenases, in *The Enzymes*, 3rd edn. (Boyer, P. D., ed.), vol. 11A, pp. 61-102, Academic Press, New York

SAMAMA, J.-P., ZEPPEZAUER, E., BIELLMANN, J.-F. & BRÄNDÉN, C.-I. (1977) The crystal structure of complexes between horse liver alcohol dehydrogenase and the coenzyme analogues 3-iodopyridine and pyridine adenine dinucleotide. *Eur. J. Biochem.* 81, 403-409

STURTEVANT, J. M. (1977) Heat capacity and entropy changes in processes involving proteins. *Proc. Natl. Acad. Sci. U.S.A.* 74, 2236-2240

THEORELL, H. & CHANCE, B. (1951) Studies on liver alcohol dehydrogenase. *Acta Chem. Scand.* 5, 1127-1144

WHITE, J. L., HACKERT, M. L., BUEHNER, M., ADAMS, M. J., FORD, G. C., LENTZ, P. J. JR., SMILEY, I. R., STEINDEL, S. J. & ROSSMANN, M. G. (1976) A comparison of the structures of apo dogfish M_4 lactate dehydrogenase and its ternary complexes. *J. Mol. Biol.* 102, 759-779

Discussion

Knowles: The position of the substrate you describe conflicts with that proposed on the basis of n.m.r. measurements by Sloan *et al.* (1975) for the cobalt-substituted enzyme: they said that the carbonyl oxygen atom of the aldehyde was not a metal ligand, but that there was a bridging water molecule. Also, they investigated the tertiary isobutyramide–NADH complex and an abortive ternary complex with the real substrates: ethanol and NADH.

Brändén: One possible explanation of the difference between our model and that of Sloan *et al.* is that we might be observing different ligands. Using n.m.r. techniques, Sloan *et al.* observe a fast-exchanging ligand which has been interpreted in terms of a second-sphere complex. With the X-ray method we observe a ligand which is firmly and directly bound to the catalytic zinc atom and which presumably exchanges slowly. Our ligand would be invisible by the n.m.r. technique and the fast-exchanging ligand might not be seen in the X-ray map.

There is, however, not only a conflict between our proposed models but also a serious discrepancy between our data. In their study of the fully substituted enzyme they computed the distances of the bound ligand from both the structural and catalytic cobalt atom. For the methyne proton of bound iso-butyramide these distances are 8.6 and 6.6 Å. The sum of these distances (15.2 Å) is considerably shorter than the distance of 20 Å between the catalytic and structural zinc atoms obtained in our X-ray structure. Clearly, their computed distances, which are based on some severe approximations, are not compatible with the X-ray structure.

Vallee: Sloan *et al.* prepared an alcohol dehydrogenase derivative containing zinc and cobalt by dialysing the native enzyme against cobalt, according to the procedures of Young & Wang (1971). In both these instances the procedures do not replace metal atoms selectively. As Sloan *et al.* pointed out, only about 80% of the cobalt was present at the rapidly-exchanging (i.e. non-catalytic) sites, and the rest was at the slowly-exchanging sites. Consequently one cannot calculate distances unambiguously and any conclusions based on such calculations must be in error.

Bränden: That is why I did not mention their work on the hybrid enzyme.

Vallee: But it is essential to the question.

Lipscomb: Did they really displace the zinc with cobalt?

Vallee: Yes, but that is not the issue; the important thing is knowing at which metal site the replacement occurred: Sloan *et al.* never demonstrated a unique exchange at either the catalytic or the structural site. The specificity of metal substitution is extremely sensitive to the exchange conditions, and this has been documented in detail by Sytkowski & Vallee (1976) who did accomplish such selective replacement.

Lipscomb: Has Mildvan ever studied your substrate bound to the cobalt enzyme?

Bränden: No.

Arigoni: Professor Bränden, is there any water in the hydrophobic barrel to begin with? If so, what happens as the substrate enters? The water must leave somehow.

Bränden: Unless nature prefers a vacuum there must be water in this barrel! The conformational change induced by binding of coenzyme closes the entrance to the active site from the front of the molecule (see Fig. 4) but not sufficiently, I am sure, to prevent water molecules passing between the two domains. Substrate molecules containing more than a couple of carbon atoms cannot pass. This question will be answered in detail when we have completed high-resolution studies, in which we make accessibility calculations of the sort that Professor Richards discussed, and can see whether there is a hole big enough for water to pass through.

Arigoni: Professor Conforth has published chemical evidence about which of the two diastereotopic sites are responsible for hydride transfer. Do you take this into account when you build the model or can you derive the specificity from the X-ray work without chemical knowledge of what goes on?

Brändén: We have been able to deduce the specificity from the following two factors. First, the position of the carboxamide group determines the specificity. We observe electron density corresponding to the carboxamide group and can thus determine the orientation of the nicotinamide ring in our X-ray maps. Secondly, when we try to build a model with the nicotinamide in the other orientation (B-side stereospecificity), we encounter severe steric hindrance from one of the zinc ligands (the S atom of Cys-174) and the main chain of Gly-175.

Phillips: Do anions bind to the carboxy-ends of the parallel pleated sheets or to the amino-ends of the associated helices?

Brändén: In alcohol dehydrogenase anions bind close to the ends of the strands of the pleated sheets but whether that is determined by the properties of the sheets, the helices or the loops between them we do not know.

Phillips: In triosephosphate isomerase the phosphate appears to be coordinated more with free NH-groups on the loop region than with helices. Nevertheless the amino-ends of helices are potentially attractive sites for phosphate binding because they have free NH-groups available to form hydrogen bonds with the oxygen atoms.

Brändén: That raises another question: why are not cations bound in the carboxy-ends of the helices?

Vallee: Is there any arginine at the carboxy-end?

Brändén: We cannot find any sort of regularity in terms of the side-chain distribution of these anion-binding sites. I presume that this type of topological arrangement has some fundamental property.

Phillips: Are all the dipoles aligned with their positive ends directed toward the anion?

Brändén: For the helices, yes.

Franks: By anion binding do you mean that one specific anion will bind to one specific enzyme or will many anions bind to these sites in different enzymes?

Brändén: The anion-binding site in liver alcohol dehydrogenase binds various anions. A heavy-atom derivative, $Pt(CN)_4^{2-}$, which is extremely stable (with a stability constant for removal of the first cyanide group of about 10^{30}) is bound as an anion at this site; Cl^- binds well; large organic molecules, such as 8-anilino-1-naphthalenesulphonate, bind through the sulphonate group in that position. How true this is for kinases and other enzymes I do not know.

Phillips: In the 1950s when active sites were being explored by studies depending on the chemical modification of substrates and inhibitors many reckless interpretations were made because it is extremely difficult to design a modification that affects the electronic distribution within a molecule but not its geometry. Once the geometry is altered all sorts of weird things happen. The first example that strongly influenced me was the finding that conversion of the α-anomer of *N*-acetylglucosamine into the β-anomer (at the time a seemingly trivial change to biochemists) drastically affects the binding of that sugar to lysozyme (Blake *et al.* 1967). Procedures that modify the electronic properties of molecules and not their spatial ones are hard to find.

Brändén: The time may now be ripe to review this question and to do a much more careful analysis of modified coenzyme analogues with a combination of chemical and crystallographic techniques.

Lipscomb: Is the mechanism of hydride transfer from the nicotinamide ring established or is some other mechanism (such as a radical one) operating?

Brändén: Since the 1940s the mechanism has been assumed to be hydride transfer. Klinman (1976) has, however, suggested the involvement of a protonated radical intermediate in the reaction sequence of yeast alcohol dehydrogenase. If a hydrogen atom and one electron were transferred in separate steps and if the time difference between these two events were large enough, it should be possible to detect an e.s.r. signal. We have done such experiments (unpublished work) using the liver enzyme at liquid helium temperature (with rapid freezing of an equilibrium mixture) but observed no e.s.r. signal. That, of course, does not exclude the possible existence of a short-lived radical intermediate but thermodynamic calculations (Blankenhorn 1976) have shown that such a mechanism is unfavourable owing to the high energy of the nicotinamide radical.

References

BLAKE, C. C. F., JOHNSON, L. N., MAIN, G. A., NORTH, A. C. T., PHILLIPS, D. C. & SARMA, V. R. (1967) Crystallographic studies of the activity of hen egg-white lysozyme. *Proc. R. Soc. Lond. B Biol. Sci. 167*, 378-388

BLANKENHORN, G. (1976) Nicotinamide-dependent one-electron and two-electron (flavin) oxidoreduction: thermodynamics, kinetics and mechanism. *Eur. J. Biochem. 67*, 78-80

KLINMAN, J. P. (1976) Isotope effects and structure–reactivity correlations in the yeast alcohol dehydrogenase reaction. *Biochemistry 15*, 2018-2026

SLOAN, D. L., YOUNG, J. M. & MILDVAN, A. S. (1975) Nuclear magnetic resonance studies of substrate interaction with cobalt substituted alcohol dehydrogenase from liver. *Biochemistry 14,* 1998-2008

SYTKOWSKI, A. J. & VALLEE, B. L. (1976) Chemical reactivities of the catalytic and non-catalytic zinc or cobalt atoms of horse liver alcohol dehydrogenase: differentiation by their thermodynamic and kinetic properties. *Proc. Natl. Acad. Sci. U.S.A. 73,* 344-348

YOUNG, J. M. & WANG, J. H. (1971) The nature of binding of competitive inhibitors to a alcohol dehydrogenase. *J. Biol. Chem. 246,* 2815-2821

The consequences of nucleotide binding to liver alcohol dehydrogenase

H. GUTFREUND

Department of Biochemistry, University of Bristol

Abstract   Extensive and informative steady-state kinetic investigations of the mechanisms of horse liver alcohol dehydrogenase have recently been complemented by observations of the fluorescence and spectroscopic characteristics of transient intermediates by rapid-reaction techniques. In this way it was possible to study separately steps involved during enzyme–substrate complex formation and during the catalytic process.

It can be shown that a proton is liberated during complex formation before the transfer of a hydride ion from ethanol to form NADH. This must be due to a change in pK of a group on the enzyme protein and is linked to a change in tryptophan fluorescence.

Pressure relaxation techniques have enabled us to study the rate constants of the change in tryptophan fluorescence linked to NAD^+ binding and proton dissociation. We have shown that NAD^+ binding occurs in two steps: a rapid second-order association, followed by the substrate–induced isomerization to form the reactive enzyme–substrate intermediate. The isomerization rate constants were determined in both directions and their role in the overall reaction mechanism could be identified.

Liver alcohol dehydrogenase (EC 1.1.1.1) played an important part in the development of methods for the investigation of the kinetic mechanisms of enzymes in general and NAD^+-linked dehydrogenases in particular. Shortly after its isolation and crystallization Theorell & Chance (1951) studied this enzyme by transient-kinetic techniques, which had previously only been applied to haem proteins. The application of steady-state kinetic analysis was also advanced considerably when Dalziel (1958) developed a new approach using liver alcohol dehydrogenase as an example. The model proposed (the Theorell–Chance mechanism) was based on the observation that the ternary complexes (enzyme–nucleotide–substrate/product) are not kinetically significant; that is, their concentrations are not appreciable during steady-state turnover and the rates of interconversion are not observed.

81

For the understanding of the mechanism of an enzyme, it is important to get kinetic information about the catalytic step (interconversion of enzyme–substrate into enzyme–product complexes) as well as about the non-covalent steps during formation and decomposition of these complexes. Shore & I have applied the kinetic analysis of the transient approach to the steady state, developed for the study of hydrolytic enzymes (Gutfreund 1965). Dalziel (1975) has summarized the complementary information available from steady-state and transient-kinetic studies.

I shall review briefly the findings of transient-kinetic investigations and present the results of some experiments with relaxation techniques, which might be extended to obtain information about the mechanism of liver alcohol dehydrogenase.

TRANSIENT NADH PRODUCTION

Shore & I (1970) showed that the catalytic interconversion of the ternary complexes, in the direction of alcohol oxidation, can be observed during the transient approach to the steady state as the formation of enzyme-bound NADH. The steady-state rate is controlled by the slow dissociation of NADH from the enzyme; aldehyde dissociates much more rapidly. The fact that the transient formation of one mole of NADH per enzyme active site is at least principally controlled by the chemical step of hydride transfer is documented by a large deuterium isotope effect and the different rates with different alcohols. The extension of studies with different alcohols will be of considerable interest for the interpretation of the chemical reaction mechanism when the catalytic groups on the enzyme are defined by the combined crystallographic, chemical and kinetic studies in progress (Parker et al. 1978; Brändén, this volume).

TRANSIENT H+ FORMATION

Steady-state analysis and equilibrium binding of NAD+ indicated that complex formation of the enzyme with this nucleotide results in the shift of the pK of a group on the protein by at least two units to the acid side (see Dalziel 1963). Transient-kinetic observation of the formation of enzyme-bound NADH and of the liberation of protons gave the following results. With ethanol and NAD+ as substrates the formation of one equivalent of H+ was slightly faster than that of the first equivalent of NADH per active site. With [$^2H_5$]ethanol the transient formation of NADH was slowed down by a factor of four but the rate of the proton transient was not affected. Subsequent experiments (Shore et al. 1975) showed that the protein fluorescence changed during the formation of the enzyme–substrate complex. The equilibrium relationships between this substrate-induced transformation, the shift in pK and the pH dependence of protein fluorescence were further analysed (Wolfe et al. 1977).

THE SUBSTRATE-INDUCED CONFORMATION CHANGE

The kinetics of the fluorescence change and the linked change in the enzyme–NAD$^+$ binding equilibrium on protonation of the enzyme–NAD$^+$ complex proved itself to be amenable to investigation by the pressure-relaxation method (Davis & Gutfreund 1977). Experiments with enzyme in the absence of substrate showed that a rapid conformational transition occurred with a relaxation time of less than 100 μs at alkaline pH.

The relaxations of the enzyme–NAD$^+$ system after pressure perturbation were investigated in detail as a function of NAD$^+$ concentration and pH (Coates et al. 1977). The principal relaxation observed in these studies could be interpreted in terms of the following mechanism:

$$\text{HE} + \text{NAD}^+ \overset{k_1}{\rightleftharpoons} \text{HENAD}^+ \overset{k_2}{\rightleftharpoons} \text{H}\overset{*}{\text{E}}\text{NAD}^+ \overset{k_3}{\rightleftharpoons} {}^-\overset{*}{\text{E}}\text{NAD}^+ + \text{H}^+$$

The reciprocal relaxation time for this mechanism, assuming rapid equilibration of steps 1 and 3, is given by equation (1) where [HE] and [NAD$^+$] are

$$\tau^{-1} = \frac{k_2}{1 + \{k_{-1}/k_1 ([\text{HE}] + [\text{NAD}^+])\}} + \frac{k_{-2}}{1 + (K_A/[\text{H}^+])} \tag{1}$$

the equilibrium concentrations of the protonated enzyme and NAD$^+$, respectively, and K_A is the acid dissociation constant of the species HENAD$^+$. The results are in agreement with $K_A = 10^{-7.6}$ proposed by Dalziel (1963). At pH 7.7 and saturating NAD$^+$ concentration $\tau^{-1} = 360$ s^{-1}. In the presence of trifluoroethanol the relaxation was considerably faster, approaching that of the free enzyme.

DISCUSSION OF RESULTS

The binding of NAD$^+$ to horse liver alcohol dehydrogenase results in a substrate-induced isomerization which quenches the fluorescence of one of the two tryptophyl groups of the enzyme subunit and causes the release of a proton at neutral pH. When both NAD$^+$ and alcohol are bound, this isomerization is much faster than either the catalytic step or release of product. However, when the reaction proceeds in the other direction (reduction of aldehyde) and alcohol, the first product to dissociate, is present at low concentration, the isomerization will control NAD^+ dissociation.

The system of groups at the active site of liver alcohol dehydrogenase, which is responsible for the cyclic release and uptake of a proton during enzyme turnover, is described in the discussion following this paper. In addition to the information on the proton release linked to the NAD^+-induced conformational change, one has to take into account the experiments of Brooks et al. (1972) which showed that a group with pK about 6.5 has to be in the basic form for the catalytic step to proceed.

The sequence of events proposed for alcohol dehydrogenase is somewhat different than that proposed by Holbrook & Gutfreund (1975) for the reaction of lactate dehydrogenase. During the turnover of the latter enzyme the proton is abstracted from the OH group of the substrate by the imidazole of His-51 and proton release is linked to substrate dissociation.

References

BROOKS, R. L., SHORE, J. D. & GUTFREUND, H. (1972) The effects of pH and temperature on hydrogen transfer in the liver alcohol dehydrogenase mechanism. *J. Biol. Chem. 247*, 2382-2383

COATES, J. H., HARDMAN, M. J., SHORE, J. D. & GUTFREUND, H. (1977) Pressure relaxation studies of isomerisations of horse liver alcohol dehydrogenase linked to NAD^+ binding. *FEBS (Fed. Eur. Biochem. Soc.) Lett. 84*, 25-28

DALZIEL, K. (1958) The determination of constants in a general coenzyme reaction mechanism by initial rate measurements in the steady state. *Trans. Farad. Soc. 54*, 1247-1253

DALZIEL, K. (1963) Kinetic studies of liver alcohol dehydrogenase and pH effect with coenzyme preparations of high purity. *J. Biol. Chem. 238*, 2850-2858

DALZIEL, K. (1975) Dynamic aspects of enzyme specificity. *Philos. Trans. R. Soc. Lond. Biol. Sci. 272*, 109-122

DAVIS, J. S. & GUTFREUND, H. (1977) The scope of moderate pressure changes for kinetic and equilibrium studies of biochemical systems. *FEBS (Fed. Eur. Biochem. Soc.) Lett. 72*, 199-207

GUTFREUND, H. (1965) *An Introduction to the Study of Enzymes*, pp. 54-61 and 255-262, Blackwell, Oxford

HOLBROOK, J. J. & GUTFREUND, H. (1975) Approaches to the study of enzyme mechanisms: lactate dehydrogenase. *FEBS (Fed. Eur. Biochem. Soc.) Lett. 3!*, 157-169

PARKER, D. M., HARDMAN, M. J., PLAPP, B. V. & HOLBROOK, J. J. (1978) pH dependent changes of intrinsic fluorescence of chemically modified liver alcohol dehydrogenase. *Biochem. J.*, in press

SHORE, J. D. & GUTFREUND, H. (1970) Transients in the reactions of liver alcohol dehydrogenase. *Biochemistry 9*, 4655-4659

SHORE, J. D., GUTFREUND, H. & YATES, D. W. (1975) Quenching of protein fluorescence by transient intermediates in the liver alcohol dehydrogenase reaction. *J. Biol. Chem. 250*, 5276-5277

THEORELL, H. & CHANCE, B. (1951) Studies on liver alcohol dehydrogenase II. The kinetics of compound of horse liver alcohol dehydrogenase and reduced diphosphopyridine nucleotide. *Acta Chem. Scand. 5*, 1127-1144

WOLFE, J. K., WEIDIG, C. F., HALVORSON, H. R., SHORE, J. D., PARKER, D. M. & HOLBROOK, J. J. (1977) pH dependent conformational states of horse liver alcohol dehydrogenase. *J. Biol. Chem.* *252*, 433-436

Discussion

Brändén: If these pK_a changes are not associated with groups on the surface of the molecule affected by the coenzyme-induced conformational change but reflect properties of groups directly involved in catalysis, there are few possible candidates. Apart from the zinc ligands and the zinc-bound water molecule the only groups available are Ser-48 and His-51. In the vicinity of the active site there is no other group from which a proton could be released or which could accept the proton.

Williams: Has the sulphur atom been eliminated as a possible acceptor for the proton?

Brändén: As I said before, among the groups available are the zinc ligands which includes the sulphur atoms of Cys-46 and Cys-174.

Williams: The sulphur atom of Cys-46 is extremely reactive and open to attack.

Vallee: It reacts with iodoacetate owing to the presence of Arg-47 which attracts the carboxy group of the iodoacetate. The thiol is converted into a thioether but still serves as a zinc ligand. The iodide coordinates to the zinc atom (as the X-ray data show). On the other hand, the reaction of Cys-46 with iodoacetamide is much less, presumably because it cannot bind to Arg-47.

Brändén: May I moderate Professor Williams' statement? First, Cys-46 is not particularly reactive as reactive thiol groups go; the essential SH group in glyceraldehydephosphate dehydrogenase is far more reactive towards iodoacetate. Secondly, two SH groups in the yeast enzyme are modified to the same extent by iodoacetamide (Twu *et al.* 1973). These two residues correspond to Cys-46 and Cys-174 in the liver enzyme. Furthermore, several other reagents have been tested (Brändén *et al.* 1975) which show that, depending on the direction of the attack, these two cysteinyl groups have similar reactivities.

Vallee: In this context iodoacetate probably acts as an active-site-directed reagent because of its interaction with the arginyl residue.

Williams: But do you exclude proton attack not at the seryl OH but at the sulphur atom on the metal?

Vallee: That is possible.

Battersby: How are the ligands arranged around the zinc atom?

Brändén: As far as we can tell, they are arranged in a distorted tetrahedron.

Vallee: Just as they should be!

Brändén: There is space available for a fifth ligand — a water molecule — for those who want to include acid-base catalysis in this model.

Cornforth: Brooks & Shore (1971) showed a deuterium isotope effect of 4.3 for MeOH on liver alcohol dehydrogenase. I assume that that was the difference in velocity between methanol and $[^2H_3]$MeOH. Has any experiment been done to show the true intramolecular deuterium effect in that case, that is to say, to what extent a partially deuteriated methanol shows intramolecular discrimination between hydrogen and deuterium?

Gutfreund: The oxidation of methanol is slower than the dissociation of NADH. With this substrate one is looking at the steady-state rather than at the transient. As we have not done more detailed experiments with partially labelled methanol I cannot answer your question.

Arigoni: What factor is responsible for stabilizing the positive charge which develops on the pyridinium nitrogen atom as the enzyme proceeds from the reduced form, after hydride transfer, to the oxidized form (see 1)? Is there a

counterion to stabilize the charge that is being developed on the pyridinium ion that is formed? Is there any relay between the upper and lower parts of the system?

Brändén: In the mechanism we have proposed (Brändén *et al.* 1975) the stabilizing negative charge is provided by the negatively charged alcoholate ion in the ternary enzyme–NAD+–alcohol complex. After dissociation of the alcohol, this charge is provided by a zinc-bound hydroxy ion which accepts the proton that is taken up on dissociation of NAD+. The corresponding complexes with the reduced form of the coenzyme contain the uncharged aldehyde or water molecules. In view of the recent results presented by Professor Gutfreund this may be a simplification of a more complex system involving Ser-48 and His-51 and possibly also another zinc-bound water molecule as a fifth ligand as suggested by Dworschack & Plapp (1977).

Knowles: Adams *et al.* (1973) said that lactate dehydrogenase had a glutamic acid residue nicely placed for the stabilization of the developing positive charge on the pyridine-ring nitrogen atom (as NADH goes to NAD+) that Professor Arigoni described. This prompted Hajdu & Sigman (1977) to do several studies on NADH models with neighbouring carboxylate ions, the results of which suggested that there was (at least in the models) some electrostatic catalysis of the kind that Professor Lipscomb has summarized (elsewhere in this volume). Unfortunately, however, the complete sequence of lactate dehydrogenase (Taylor 1977) now shows that there is *no* acid anion near to that nitrogen atom. Which shows how nature does not always use devices that may seem attractive to chemists.

Brändén: I agree. The sequence of the lactate dehydrogenases from horse and rat have been independently determined (Jörnvall 1974) and match well; furthermore, we have found no conflict between sequence or shape of sidechains in the electron-density maps. There is definitely no negatively charged residue in the vicinity of the nitrogen atom of the nicotinamide group. The presence of such a charge would be in strong conflict with the observation that NADH is bound one order of magnitude more tightly than NAD$^+$ at neutral pH both for lactate and alcohol dehydrogenase.

Vallee: I might just mention that various anions (chloride, pyrophosphate, formate) affect the interaction between liver alcohol dehydrogenase and NADH (Li *et al.* 1963). The reason has never been established, but they do decrease coenzyme binding.

Gutfreund: By how much?

Vallee: NADH binding is reduced by about 50% in 0.5m-chloride and even more in the same concentration of formate (Ulmer & Vallee 1965).

References

ADAMS, J. M., BUECHNER, M., CHANDVASEKHAR, K., FORD, G. C., HACKERT, M. L., LILJAS, A., ROSSMAN, M. G., SMILEY, I. E., ALLISON, W. S., EVERSE, J., KAPLAN, N. O. & TAYLOR, S. S. (1973) Structure-function relationships in lactate dehydrogenase. *Proc. Natl. Acad. Sci. U.S.A.* 70, 1968-1972

BRÄNDÉN, C.-I., JÖRNVALL, H., EKLUND, H. & FURUGREN, B. (1975) Alcohol dehydrogenases, in *The Enzymes*, 3rd edn., vol. XIA (Boyer, P.D., ed.), pp. 103-190, Academic Press, New York

BROOKS, R. L. & SHORE, J. D. (1971) Effect of substrate structure on the rate of the catalytic step in the liver alcohol dehydrogenase mechanism. *Biochemistry 10*, 3855-3857

DWORSCHACK, R. T. & PLAPP, B. V. (1977) pH, Isotope and substituent effects on the interconversion of aromatic substrates catalyzed by hydroxybutyrimidylated liver alcohol dehydrogenase. *Biochemistry 16*, 2716-2725

HAJDU, J. & SIGMAN, D. S. (1977) Model dehydrogenase reactions. Catalysis of dihydronicotinamide reductions by noncovalent interactions. *Biochemistry 16*, 2841-2846

JÖRNVALL, H. (1974) Functional aspects of structural studies of alcohol dehydrogenases, in *Alcohol and Aldehyde Metabolizing Systems* (Thurman, R., ed.), pp. 23-32, Academic Press, New York

LI, I.-K., ULMER, D. D. & VALLEE, B. L. (1963) Anomalous rotatory dispersion of enzyme complexes. IV. Mechanism of inhibition of liver alcohol dehydrogenase by buffer anions and bases. *Biochemistry 3*, 482-486

TAYLOR, S. S. (1977) Amino acid sequence of dogfish muscle lactate dehydrogenase. *J. Biol. Chem.* 252, 1799-1806

TWU, J., CHIN, C. C. Q. & WOLD, F. (1973) Studies on the active-site sulfhydryl groups of yeast alcohol dehydrogenase. *Biochemistry 12,* 2856-2862

ULMER, D. D. & VALLEE, B. L. (1965) Extrinsic Cotton effects and the mechanism of enzyme action. *Adv. Enzymol. 27,* 37-104

Origins of specificity in the binding of small molecules to dihydrofolate reductase

GORDON C. K. ROBERTS

Division of Molecular Pharmacology, National Institute for Medical Research, London

Abstract   Dihydrofolate reductase is the target for the therapeutically important 'anti-folate' drugs such as methotrexate and trimethoprim. Methotrexate is a powerful inhibitor of the enzyme, binding up to 10 000 times more tightly than the structurally similar substrate, folate. Two contributions to this striking difference in affinity have been identified: the two ligands bind in different charge states, and there are conformational differences between the two complexes.

The origins of the tight binding of methotrexate have been explored further by studying the binding of 2,4-diaminopyrimidine and *p*-aminobenzoyl-L-glutamate, which may be considered as 'fragments' of methotrexate. These two compounds bind simultaneously but also cooperatively, the binding of one 'fragment' leading to a 50-fold increase in the affinity for the other. Studies of structural analogues of these fragments show that the specificity as well as the strength of binding can be altered by the presence of the other 'fragment'; both positive and negative cooperativity are observed. The relation of these observations to methotrexate binding, and the notion of intramolecular cooperativity in ligand binding are discussed.

One of the most striking and biologically most important attributes of proteins is the specificity they show in their interactions with small molecules. This specificity arises from the interplay of many weak intermolecular forces. When a crystallographic picture of the protein–small molecule complex is available, several of these individual interactions (such as hydrogen bonds and charge–charge interactions) can often be identified, and some qualitative statements about the origins of specificity can be made. However, it is much more difficult to estimate the relative importance of the different interactions in the overall binding process. It is probably fair to say that there is no single case in which we can explain *quantitatively* why one small molecule binds, say, 100-fold more tightly than another to a particular protein. The problem is one of relating structure to energetics. To solve it we need a better understanding not only of intermolecular forces in aqueous solution but also of dynamic

89

aspects of complex formation (see King & Burgen 1976) and in particular of the extent to which the structure of the binding site can change so as to accommodate itself to ligands of different structure.

Aside from its intrinsic interest, a solution to this problem would be of some practical importance, since it might be expected to lead to advances in the rational design of drugs. An enzyme of particular interest in this regard is dihydrofolate reductase, which catalyses the NADPH-linked reduction of dihydrofolate to tetrahydrofolate. Tetrahydrofolate is vital in intermediary metabolism, acting as a 'carrier' of one-carbon fragments in the biosynthesis of amino-acids, thymidylate and purines (for a review see Blakley 1969). In most of these one-carbon transfers, the folate coenzyme remains at the tetra-hydro oxidation level, but in the synthesis of thymidylate from deoxyuridylate (catalysed by thymidylate synthetase) the product is dihydrofolate. This must then be reduced by dihydrofolate reductase to maintain the cellular pool of tetrahydrofolate derivatives. Inhibition of dihydrofolate reductase will thus lead to a depletion of this pool and to inhibition of DNA synthesis, owing to a lack of thymidylate and purines. Many potent inhibitors of dihydrofolate reductase have been synthesized and several have found clinical application, notably in the treatment of bacterial infections (trimethoprim), malaria (pyrimethamine) and some neoplastic diseases (methotrexate). (For reviews see Hitchings & Burchall 1965; Baker 1967; Blakley 1969; Bertino 1971).

The structures of the substrates dihydrofolate and folate are compared with those of the inhibitors methotrexate and trimethoprim in Fig. 1. Although the substrates are 2-amino-4-oxo substituted pteridines, all the most effective inhibitors have 2,4-diamino substituents. Thus methotrexate differs from folate essentially only in the replacement of the 4-oxo group by a 4-amino group (methylation on N-10 has little effect on the interaction with the enzyme), yet it is not a substrate and binds as much as 10 000 times more tightly to the enzyme than does folate. The essential structural fragment seems to be a 2,4-diaminopyrimidine ring, as in trimethoprim. This compound illustrates another striking aspect of the specificity of this enzyme: trimetho-prim binds some 30 000-fold more tightly to the bacterial enzyme than to mammalian dihydrofolate reductase – hence its effectiveness in the treatment of bacterial infections. The species specificity arises from the trimethoxyben-zyl group, which makes a major contribution to the binding of trimethoprim to the bacterial enzyme but which has no counterpart in the structure of the natural substrate. Thus the structure–activity relationships among substrates and inhibitors of dihydrofolate reductase show several features which are both of practical importance and a challenge to our understanding of the origins of specificity in the binding of small molecules to proteins.

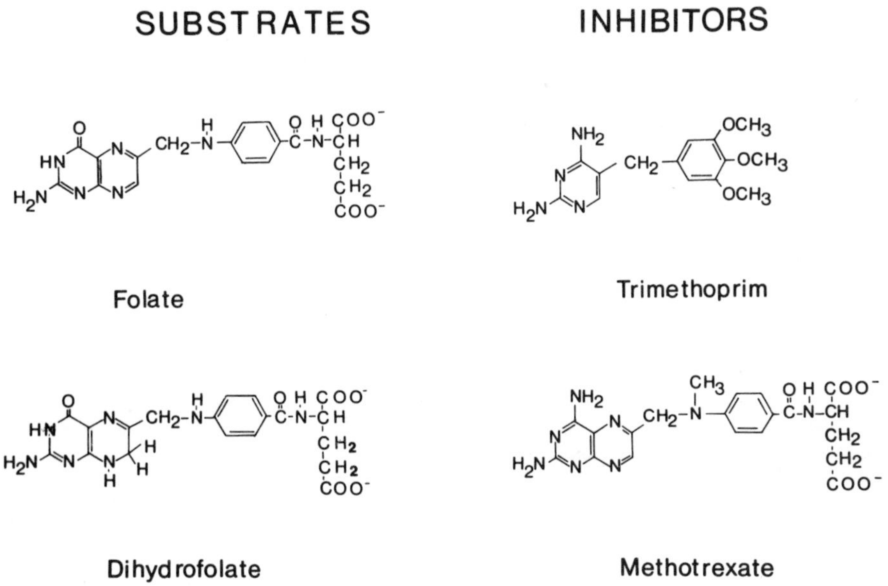

SUBSTRATES

INHIBITORS

Folate

Trimethoprim

Dihydrofolate

Methotrexate

FIG. 1. Structures of the substrates and of two important inhibitors of dihydrofolate reductase.

In an attempt to throw some light on the molecular basis of these effects, we have been studying the binding of substrates and inhibitors to *Lactobacillus casei* dihydrofolate reductase by various spectroscopic and kinetic methods. I shall concentrate here on some recent experiments which illustrate possible approaches to a quantitative understanding of specificity, and also some of the difficulties which arise.

CHARGE–CHARGE INTERACTIONS IN METHOTREXATE BINDING

Some years ago Baker (see Baker 1967) and Pullman (Perault & Pullman 1961; Collin & Pullman 1964) drew attention to the increased basicity of the N-1 atom of the pteridine ring which results from replacing the 4-oxo substituent by an amino group. Baker suggested that the 2,4-diamino compounds (such as methotrexate) bind to the enzyme in the protonated form and that their tighter binding can be explained by an additional charge–charge interaction with a group on the enzyme which is not possible for 2-amino-4-oxo compounds such as folate. The recent determination of the crystal structure of the complex between *Escherichia coli* dihydrofolate reductase and methotrexate (Matthews *et al.* 1977) reveals that the carboxylate groups of Asp-27 (a residue conserved in all dihydrofolate reductases of known sequence) is

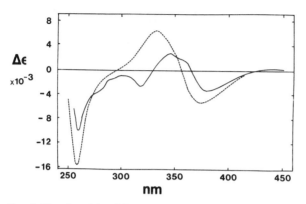

FIG. 2. The ultraviolet difference spectrum generated on methotrexate binding to *L. casei* dihydrofolate reductase (solid line), compared with that generated on protonation of free methotrexate (dashed line) (from Hood & Roberts 1978).

in close proximity to the N-1 atom of the bound methotrexate. This clearly supports Baker's suggestion that protonated methotrexate makes a charge–charge interaction with the enzyme. However, it does not allow us to estimate the extent to which this interaction can account for the tighter binding of methotrexate as compared to folate. To do this we must turn to spectroscopic experiments.

It has been noted on several occasions (Erickson & Mathews 1972; Poe *et al.* 1974, 1975; Gupta *et al.* 1977) that there is a general resemblance between the ultraviolet difference spectrum generated when methotrexate binds to the enzyme and that generated on protonation of methotrexate (see Fig. 2 for the case of the *L. casei* enzyme). This qualitative resemblance suggests that methotrexate may be protonated when bound to the enzyme at neutral pH. However, it is clear from Fig. 2 that the similarity between the difference spectra produced by protonation and by binding to the enzyme is far from complete. Detailed analysis is facilitated by examination of the pH-dependence of the difference spectrum generated on methotrexate binding to the enzyme (Fig. 3; Hood & Roberts 1978). In this way we have resolved the observed difference spectrum into three components:

(*a*) a component corresponding closely in shape to the protonation difference spectrum, that is with negative bands at 260 nm and 375 nm and a positive band at 334 nm. This component, unlike the others, shows a marked pH-dependence of its amplitude (cf. Fig. 3);

(*b*) a component arising from a substantial red shift of one absorption band of methotrexate on binding, resulting in a negative band at 321 nm and a positive band at 362 nm and

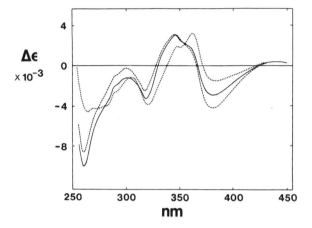

FIG. 3. The ultraviolet difference spectra generated on methotrexate binding to *L. casei* dihydro-folate reductase at various pH values: pH 5.30 (dotted line), pH 7.30 (dashed line), and pH 8.20 (solid line) (from Hood & Roberts 1978).

(*c*) a component consisting of two small positive bands at 289 and 296 nm, which most probably arises from perturbation of a tryptophyl residue when methotrexate binds (for which there is also evidence from nuclear magnetic resonance spectroscopy; Kimber *et al.* 1977).

Further details of the analysis are reported elsewhere (Hood & Roberts 1978). The demonstration of the existence of a component of the difference spectrum generated on binding which clearly corresponds to the protonation difference spectrum shows unequivocally that methotrexate exhibits a greater degree of protonation (a higher pK) in its complex with the enzyme than in free solution. This can be quantitated by analysis of the pH-dependence of the amplitude of the protonation component. Fig. 4 shows the amplitude of the observed difference spectrum at 400 nm as a function of pH; at this wavelength only the protonation component makes a significant contribution. The form of this pH-dependence is governed by the pK_a of methotrexate in the free and bound states, and by the extinction coefficients of the protonated and neutral forms in the free and bound states. Of these six parameters, three (those characterizing free methotrexate) can be measured directly, and the remaining three can be estimated by non-linear least-squares fitting of the data (Hood & Roberts 1978). The curve in Fig. 4 is the 'best-fit' theoretical curve, obtained with a value for the pK_a of bound methotrexate of 8.55 ($\pm$ 0.1). The pK_a of free methotrexate is 5.35; the increase of 3.2 units in the pK on binding shows that the protonated form of methotrexate binds 1600 ($\pm$ 300) times more tightly than does the neutral form, a difference in binding energy

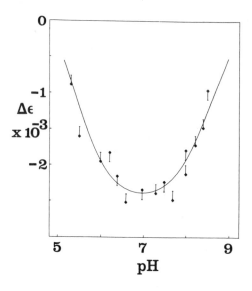

Fig. 4. The variation of the amplitude (at 400 nm) of the difference spectrum generated on methotrexate binding to dihydrofolate reductase with pH. The points are experimental, and the line is the 'best-fit' theoretical curve (from Hood & Roberts 1978).

of 4.35 ($\pm$ 0.2) kcal/mol (18.2 $\pm$ 0.8 kJ/mol). We can now assess the extent to which this increase in binding energy on protonation contributes to the difference in binding energy between methotrexate and folate. Analysis of the ultraviolet difference spectrum generated on the binding of folate to the enzyme (Hood & Roberts 1978) shows that, unlike methotrexate, folate binds as the neutral molecule, thereby confirming Baker's (1967) proposal that 2,4-diamino- and 2-amino-4-oxo-pteridines bind in different charge states. The measured binding constants for methotrexate and folate at pH 7.5 are given in Table 1. For folate the measured binding constant represents the binding constant of the neutral molecule, and for methotrexate it is a weighted average of the binding constants of neutral and protonated methotrexate (K_0 and K_+, respectively). Having measured the pK_a values of bound and free methotrexate, we can use these to calculate K_0 and K_+ from K_{app}; these values are also given in Table 1. The ratio of the apparent binding constants of methotrexate and folate at pH 7.5 is 2100 (corresponding to a Gibbs energy difference of 4.5 kcal/mol [18.8 kJ/mol]). If Baker's (1967) proposal is the whole explanation of this difference, one would expect to find that the binding constant of neutral methotrexate would be similar to that of (neutral) folate. In fact, the ratio of these two binding constants is 180, and two-thirds of the total difference in binding energy between methotrexate and folate at this pH

TABLE 1

Binding constants of folate and methotrexate to *L. casei* dihydrofolate reductase

Binding constant (l/mol)	Methotrexate	Folate
K_{app}	$2.1 \, (\pm 0.3) \times 10^8$	$9.8 \, (\pm 2.0) \times 10^4$
K_+	$2.74 \, (\pm 0.4) \times 10^{10}$	
K_0	$1.75 \, (\pm 0.6) \times 10^7$	(9.8×10^4)

Values of K_{app} were determined fluorimetrically in 100 mM-KCl, 50 mM-Tris–HCl (pH 7.5) at 25 °C. K_+ and K_0 were calculated from K_{app} and the pK_a values of methotrexate in the free and bound state (Hood & Roberts 1978).

arises from the difference between the two *neutral* molecules (Hood & Roberts 1978). Although the contribution of protonation will obviously be greater at lower pH values, this observation clearly indicates that the additional charge–charge interaction available to methotrexate is not sufficient to explain why it binds so much more tightly than does folate.

Some further contribution to the difference in binding energy may come from the different hydrogen-bonding characteristics of the two 4-substituents, although Matthews *et al.* (1977) suggest on crystallographic evidence that this will be small. There is increasing evidence that a substantial contribution to the difference in binding arises from the fact that folate and methotrexate produce different conformational changes on binding. Nuclear magnetic resonance studies (Birdsall *et al.* 1977; Feeney *et al.* 1977; Kimber *et al.* 1977; Roberts *et al.* 1977) have shown that the effects of folate and methotrexate binding on the enzyme are distinctly different. Studies of the kinetics of the binding process (R. W. King, J. G. Batchelor, K. Hood, S. Dunn & A. S. V. Burgen, unpublished work) have revealed the existence of a slow step, corresponding to a conformational change of the complex, whose equilibrium constant is an order of magnitude greater for the methotrexate than for the folate complex. A detailed structural description of these conformational changes must await crystallographic studies of the free enzyme and the enzyme–folate complex for comparison with the enzyme–methotrexate complex studied by Matthews *et al.* (1977). The available n.m.r. evidence (discussed by Roberts *et al.* 1977) suggests in a qualitative way that the differences between the folate and methotrexate complexes are extensive — in the sense that they involve many amino-acid residues — but not necessarily large.

THE BINDING OF 'FRAGMENTS' OF METHOTREXATE

In the course of these n.m.r. studies of ligand binding, we examined the

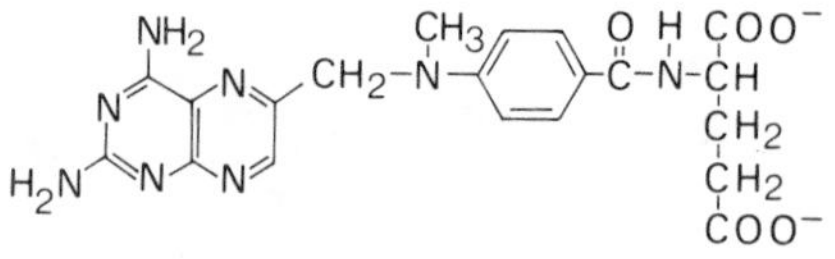

Methotrexate

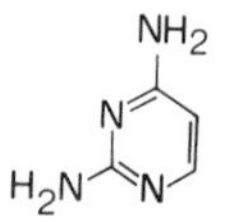

2,4-Diamino-   p-Aminobenzoyl-
  pyrimidine   L-glutamate

FIG. 5. Structures of methotrexate and its 'fragments'.

binding of 2,4-diaminopyrimidine and N-(p-aminobenzoyl)-L-glutamate, which can be regarded as 'fragments' of methotrexate (see Fig. 5). These two compounds bind cooperatively to the enzyme, in that N-(p-aminobenzoyl)-L-glutamate binds some 50 times more tightly to the enzyme–2,4-diamino-pyrimidine complex than to the enzyme alone (Table 2; Birdsall *et al.* 1978). The phenomenon of cooperativity or interdependence in the binding of two ligands to a protein is well known; it has recently been discussed from the thermodynamic point of view by Weber (1975). The present case is perhaps of particular interest since a rather large 'free-energy coupling' (Weber 1975) of −9.8 kJ/mol is observed for ligands binding to a small monomeric enzyme (molecular weight 18 000; Dann *et al.* 1976), and, as discussed later, the cooperativity is between *sub-sites* of the inhibitor binding site.

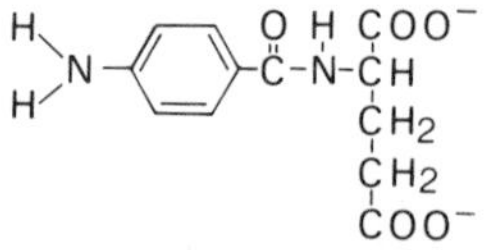

To explore the way in which this cooperativity depends on the structure of the ligands, we have investigated the binding of a series of N-(p-alkylaminobenzoyl)-L-glutamates; the results are summarized in Fig. 6 in the form of a plot of the Gibbs energy change against the alkyl chain length. The alkyl chain has a substantial effect on the binding of these compounds to the enzyme; the N-hexyl compound binds 130 times more tightly than does N-(p-aminobenzoyl)-L-glutamate itself. As can be seen from Fig. 6, the major part of this increase (27-fold) occurs on going from the N-methyl to the N-ethyl derivative. There is thus an increase in binding energy of about 8 kJ/mol on addition of a single methylene group, but addition of four more methylenes (to give the N-hexyl compound) provides only 3.3. kJ/mol more

TABLE 2

Binding constants of N-(p-aminobenzoyl)-L-glutamate and 2,4-diaminopyrimidine to *L. casei* dihydrofolate reductase

Ligand	Binary complex $K_B/(l/mol)$	Ternary complex $K_T/(l/mol)$	Cooperativity K_T/K_B
N-(p-Aminobenzoyl)-L-glutamate	$0.83\,(\pm\,0.07) \times 10^3$	$5.02\,(\pm\,0.25) \times 10^4$	$60\,(\pm\,9)$
2,4-Diaminopyrimidine	$1.28\,(\pm\,0.08) \times 10^3$	$6.02\,(\pm\,0.25) \times 10^4$	$47\,(\pm\,7)$
			(mean 54)

For each 'fragment', binding constants (in 0.5M-KCl, 15mM-bis-Tris, pH 6.0, at 25°C) were determined for binding to the enzyme alone and to the complex of the enzyme with the other 'fragment' (Birdsall *et al.* 1978).

binding energy. This is in contrast to estimates of 'hydrophobic' interaction energies from the water solubilities of alkanes, alcohols and fatty acids, which show a steady increment of about 3.3 kJ mol^{-1} methylene^{-1} (Tanford 1973). Although the interaction of the N-alkyl chains with the enzyme is most probably 'hydrophobic', the apolar region of the enzyme's surface with which the interaction takes place seems to be somewhat limited in extent, and structured in such a way that there is a particularly favourable interaction with the second methylene in the chain.

Turning now to examine the binding to the enzyme–2,4-diamino-pyrimidine complex, we find that in this situation the alkyl chain has little effect (Fig. 6): the range of binding constants is only 1.8-fold. N-(p-aminobenzoyl)-L-glutamate and its N-methyl derivative bind much more strongly in the presence of 2,4-diaminopyrimidine, but the binding of the N-ethyl to N-pentyl compounds is virtually unaffected and N-(p-hexylaminobenzoyl)-L-glutamate binds more weakly — an example of *negative* cooperativity. It is apparent that 2,4-diaminopyrimidine not only affects the affinity of the enzyme for N-(p-aminobenzoyl)-L-glutamate and related compounds but also alters the structural specificity of their binding site. A similar series of effects are seen when one examines the effects of N-(p-aminobenzoyl)-L-glutamate on the binding of 5- and/or 6-substituted 2,4-diaminopyrimidines (Birdsall *et al.* 1978). The observed cooperativity can be explained by postulating the existence of two conformations of the enzyme with different affinities for these ligands. The effects on specificity then indicate that there is a difference in *specificity* as well as affinity between the two conformations (Birdsall *et al.* 1978).

The question naturally arises whether these effects observed on the binding of 'fragments' of methotrexate have any bearing on the binding of metho-trexate itself. A detailed comparison of the binding of the 'fragments' with that

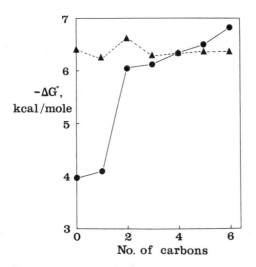

Fig. 6. The change in Gibbs energy ($-\Delta G°$) on the binding of N-(p-alkylaminobenzoyl)-L-glutamates to dihydrofolate reductase, alone (●——●) or in the presence of 2,4-diaminopyrimidine (▲---▲), as a function of the number of carbon atoms in the p-alkylamino group (from Birdsall *et al.* 1978).

of methotrexate can be made by n.m.r. spectroscopy, which allows us to monitor the behaviour of individual amino-acid residues. For dihydrofolate reductase, by preparing selectively deuteriated and fluorine-labelled analogues of the enzyme we have been able, up to the present time, to study some 25 residues in detail. As an example, the effects of the addition of 2,4-diaminopyrimidine followed by N-(p-nitrobenzoyl)-L-glutamate on the 2,6-H_2 resonances of the tyrosyl residues, observed in a selectively deuteriated enzyme, are shown in Fig. 7. For these resonances, the spectrum of the enzyme–2,4-diamino-pyrimidine–N-(p-nitrobenzoyl)-L-glutamate complex is closely similar to that of the enzyme–methotrexate complex (see Fig. 8); a similar degree of correspondence is seen in other regions of the spectrum. There is good, though indirect, evidence that some, at least, of the changes in chemical shift observed when the 'fragments' bind have their origin in a conformational change (Birdsall *et al.* 1977; Feeney *et al.* 1977; Roberts *et al.* 1977) which could well be related to the conformational change responsible for the cooperativity in ligand binding. If this is so, the close correspondence of the spectra indicated in Fig. 8 implies that the same conformational changes must accompany methotrexate binding. This in turn implies that cooperativity must be present in methotrexate binding too, though in the form of cooperativity between two parts of the same molecule — intramolecular cooperativity.

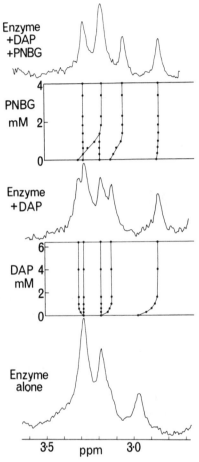

FIG. 7. The effects of adding increasing concentrations of 2,4-diaminopyrimidine (DAP), followed by *N*-(*p*-nitrobenzoyl)-L-glutamate (PNBG), on the nuclear magnetic resonance signals of the tyrosine 2,6-protons of selectively deuteriated dihydrofolate reductase (from Roberts *et al.* 1977).

GENERAL DISCUSSION

Attempts to identify the interactions responsible for the biological effects of a small molecule by relating changes in chemical structure to change in biological activity have a long and honourable history (Crum-Brown & Fraser 1868; Sexton 1963; Albert 1968). Even in the best characterized protein–small molecule systems, this remains almost the only experimental approach to the estimation of the contribution of individual interactions to the overall binding energy. Its limitations are well known — particularly the fact that no chemical change we can make will affect only one single interaction.

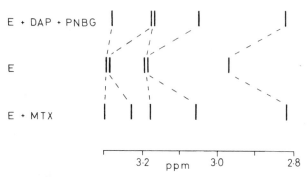

FIG. 8. A comparison of the chemical shifts of the 2,6-protons of the tyrosyl residues of dihydro-folate reductase in the free enzyme (centre), the enzyme–2,4-diaminopyrimidine–N-(p-nitrobenzoyl)-L-glutamate complex (top), and the enzyme–methotrexate complex (bottom) (from Roberts *et al.* 1977).

Thus our measurement of the change in pK of methotrexate on binding to the enzyme allows us to say that protonated methotrexate binds more tightly than the neutral molecule by 18.2 kJ/mol. However, we cannot identify this simply as the energy of the Coulombic interaction with Asp-27. The changes in electron distribution in the pteridine ring due to protonation are bound to alter the strength of interaction of other parts of the pteridine ring with the enzyme, as indeed are those due to the Coulombic interaction itself. The experimental value is only a first-order estimate of the energy of the interaction; it is to be hoped that theoretical methods will soon have advanced to the stage where we can estimate the magnitude of the effects of these electron redistributions.

These experiments demonstrate that, as Baker (1967) proposed, metho-trexate is protonated when bound to the enzyme but folate is not. However, this is not sufficient to explain the difference in binding energy between these two compounds, and a conformational difference between the complexes seems to be involved, in spite of the close similarity in structure of the two ligands. Conformational changes are also clearly implicated in the cooper-ativity in binding of the 'fragments' of methotrexate (Birdsall *et al.* 1978; Roberts *et al.* 1977). As indicated above, the similarity between the complex of the enzyme with methotrexate and that with the 'fragments' strongly suggests that *intramolecular* cooperativity is involved in methotrexate binding. This is more easily visualized if methotrexate binds in a step-wise rather than an 'all-or-none' fashion (Burgen *et al.* 1975). The notion of intramolecular cooperativity is obviously closely related to that of 'induced fit' introduced by Koshland (1958), though the latter — at least as originally formulated — refers to the effects of substrate binding in ensuring the proper orientation of *catalytic* groups, and here we are concerned with *binding*. The existence of this phenomenon clearly has important implications for the analysis of specificity. A given chemical alteration of the ligand may affect not only the interactions with the enzyme in the immediate vicinity of the modification but also the intramolecular cooperativity and hence the interactions of relatively remote parts of the ligand. This is clearly illustrated by the change in specificity of one subsite of the methotrexate binding site which accompanies ligand binding to another subsite (see Fig. 6). Once again, the effects of changing the small molecule structure cannot be localized, though here the effect is transmitted not by electron redistribution but by more or less subtle changes in the protein structure. It will be some time before theoretical methods are sufficiently developed to deal with this situation quantitatively. However, it seems likely that the use of 'fragments' of the small molecule, which suggested the existence of the phenomenon in the present case, can be extended to give at least a semiquantitative estimate of the magnitude of the effect. An essential part of this approach must be the use of spectroscopic methods, in particular n.m.r. spectroscopy, to establish the mode of binding of the various ligands so their binding energies can be properly compared. Indeed one of the most important uses of spectroscopic methods in this context is as a link between the thermodynamic, structural (crystallographic) and dynamic information which must be combined together in order to understand in detail the origins of specificity in the binding of small molecules to proteins.

ACKNOWLEDGEMENTS

I express my thanks to all the past and present members of the Division of Molecular Pharmacology who collaborated in the work described here, particularly to Arnold Burgen, Jim Feeney, Rod King, Berry Birdsall, Kevin Hood, Jaap de Miranda, Vaughan Griffiths, Gill Ostler and Barry Kimber.

References

ALBERT, A. (1968) *Selective Toxicity*, 4th edn., Methuen, London

BAKER, B. R. (1967) *Design of Active-Site-Directed Irreversible Enzyme Inhibitors*, Wiley, New York

BERTINO, J. R. (ed.) (1971) Folate antagonists as chemotherapeutic agents. *Ann. N. Y. Acad. Sci. 186*, 1-519

BIRDSALL, B., GRIFFITHS, D. V., ROBERTS, G. C. K., FEENEY, J. & BURGEN, A. S. V. (1977a) ¹H nuclear magnetic resonance studies of *Lactobacillus casei* dihydrofolate reductase: effects of substrate and inhibitor binding on the histidine residues. *Proc. R. Soc. Lond. B Biol. Sci. 196*, 251-256

BIRDSALL, B., DE MIRANDA, J. R., BURGEN, A. S. V. & ROBERTS, G. C. K. (1978) Cooperativity in ligand binding to dihydrofolate reductase. *Biochemistry 17*, 2102-2110

BLAKLEY, R. L. (1969) *The Biochemistry of Folic Acid and Related Pteridines,* North-Holland, Amsterdam

BURGEN, A. S. V., ROBERTS, G. C. K. & FEENEY, J. (1975) Binding of flexible ligands to macromolecules. *Nature (Lond.) 253,* 753-755

COLLIN, R. & PULLMAN, B. (1964) On the mechanism of folic acid reductase inhibition. *Biochim. Biophys. Acta 89,* 232-241

CRUM-BROWN, A. & FRASER, T. R. (1868) *Trans. R. Soc. Edinburgh 25,* 151

DANN, J. G., OSTLER, G., BJUR, R. A., KING, R. W., SCUDDER, P., TURNER, P. C., ROBERTS, G. C. K., BURGEN, A. S. V. & HARDING, N. G. L. (1976) Large-scale purification and characterisation of dihydrofolate reductase from a methotrexate-resistant strain of *Lactobacillus casei. Biochem. J. 157,* 559-571

ERICKSON, J. S. & MATHEWS, C. K. (1972) Spectral changes associated with binding of folate compounds to bacteriophage T4 dihydrofolate reductase. *J. Biol. Chem. 247,* 5661-5667

FEENEY, J., ROBERTS, G. C. K., BIRDSALL, B., GRIFFITHS, D. V., KING, R. W., SCUDDER, P. & BURGEN, A. S. V. (1977) ^{1}H nuclear magnetic resonance studies of the tyrosine residues of selectively deuterated *Lactobacillus casei* dihydrofolate reductase. *Proc. R. Soc. Lond. B Biol. Sci. 196,* 267-290

GUPTA, S. V., GREENFIELD, N. J., POE, M., MAKULU, D. R., WILLIAMS, M. N., MOROSON, B. A. & BERTINO, J. R. (1977) Dihydrofolate reductase from a resistant subline of the L1210 lymphoma. Purification by affinity chromatography and ultraviolet difference spectrophotometric and circular dichroic studies. *Biochemistry 16,* 3073-3079

HITCHINGS, G. H. & BURCHALL, J. J. (1965) Inhibition of folate biosynthesis and function as a basis for chemotherapy. *Adv. Enzymol. 27,* 417-482

HOOD, K. & ROBERTS, G. C. K. (1978) Ultraviolet difference spectroscopic studies of substrate and inhibitor binding to *Lactobacillus casei* dihydrofolate reductase. *Biochem. J. 171,* 357-366

KIMBER, B. J., GRIFFITHS, D. V., BIRDSALL, B., KING, R. W., SCUDDER, P., FEENEY, J., ROBERTS, G. C. K. & BURGEN, A. S. V. (1977) ^{19}F nuclear magnetic resonance studies of ligand binding to 3-fluorotyrosine- and 6-fluorotryptophan-containing dihydrofolate reductase from *Lactobacillus casei. Biochemistry 16,* 3492-3500

KING, R. W. & BURGEN, A. S. V. (1976) Kinetic aspects of structure–activity relations: the binding of sulphonamides by carbonic anhydrase. *Proc. R. Soc. Lond. B Biol. Sci. 193,* 107-125

KOSHLAND, D. E. JR. (1958) Application of a theory of enzyme specificity to protein synthesis. *Proc. Natl. Acad. Sci. U.S.A. 44,* 98-104

MATTHEWS, D. A., ALDEN, R. A., BOLIN, J. T., FREER, S. T., HAMLIN, R., XUONG, N., KRAUT, J., POE, M., WILLIAMS, M. & HOOGSTEEN, K. (1977) Dihydrofolate reductase: X-ray structure of the binary complex with methotrexate. *Science (Wash. D. C.) 197,* 452-455

PERAULT, A. M. & PULLMAN, B. (1961) Structure électronique et mode d'action des antimétabolites de l'acide folique. *Biochim. Biophys. Acta 52,* 266-280

POE, M., GREENFIELD, N. J., HIRSHFIELD, J. M. & HOOGSTEEN, K. (1974) Dihydrofolate reductase from a methotrexate-resistant strain of *Escherichia coli:* binding of several folates and pteridines as monitored by ultraviolet difference spectroscopy. *Cancer Biochem. Biophys. 1,* 7-11

POE, M., BENNETT, C. D., DONOGHUE, D., HIRSHFIELD, J. M., WILLIAMS, M. N. & HOOGSTEEN, K. (1975) Mammalian dihydrofolate reductase: porcine liver enzyme, in *Chemistry and Biology of Pteridines (Proc. 5th Intern. Symp.)* (Pfleiderer, W., ed.), pp. 51-59, Gruyter, Berlin

ROBERTS, G. C. K., FEENEY, J., BIRDSALL, B., KIMBER, B., GRIFFITHS, D. V., KING, R. W. & BURGEN, A. S. V. (1977) Dihydrofolate reductase: the use of fluorine-labelled and selectively deuterated enzyme to study substrate and inhibitor binding, in *NMR in Biology* (Dwek, R. A., Campbell, I. D., Richards, R. E. & Williams, R. J. P., eds.), pp. 95-109, Academic Press, London

SEXTON, W. A. (1963) *Chemical Constitution and Biological Activity,* 3rd edn., Spon, London

TANFORD, C. (1973) *The Hydrophobic Effect,* Academic Press, New York

WEBER, G. (1975) Energetics of ligand binding to proteins. *Adv. Protein Chem. 29,* 1-83

Discussion

Topping: The inhibitors of dihydrofolate reductase contain 4-amino groups instead of 4-oxo groups. Have the 4-methoxy compounds been investigated?

Roberts: In general, 4-methoxy compounds bind even worse than 4-oxo compounds do; I suspect that the methyl group causes some steric hindrance. A 4-amino group seems to be essential for good inhibition.

Battersby: How do you aim to assign the resonances to individual tyrosyl residues?

Roberts: Two general ways are open to us: one, sometimes known as crystal-gazing, is to wait until the crystallographers have determined the structure and then see if we can fit their information on, say, ligand binding to a specific tyrosyl residue with the n.m.r. shifts that we see. This method is, in general, extremely unsatisfactory. The second way is to use a chemical procedure, such as modification of the tyrosyl group. For dihydrofolate reductase we can modify individual tyrosyl groups by selective chemical methods, such as nitration. It is difficult to assign all the signals in this way because we have not yet found a selective handle for each residue. Consequently one has to do some crystal-gazing to complete the assignments. Usually a combination of at least two methods is needed. How one can distinguish the methyl resonances of individual valyl and isoleucyl residues (and so on) other than by crystal-gazing, I don't know.

Cornforth: What happens if you use diaminopteridine instead of diamino-pyrimidine with *p*-aminobenzoylglutamate?

Roberts: I suspect that the pteridine and *p*-aminobenzoylglutamate would not bind simultaneously; they would probably collide with one another. Unfortunately, we have not been able to do n.m.r. experiments with diaminopteridine to see if it binds in the same way as diaminopyrimidine because it is very insoluble in water.

Blundell: In the paper by Matthews *et al.* (1977) it is hard to tell whether they were working on the NADP–enzyme complex or not. To what do your data refer?

Roberts: The data I presented were all obtained in the absence of coenzyme. Matthews *et al.* studied the crystal structure of the complex of methotrexate with *E. coli* dihydrofolate reductase; coenzyme was absent. They now also have data for the ternary complex with NADPH. As with other dehydrogenases and, as we heard from Dr. Brändén, the coenzyme causes substrates and inhibitors to bind more tightly. This effect varies from a factor of about three for *p*-aminobenzoyl-L-glutamate up to about 100-fold for methotrexate. However, coenzyme does not produce any qualitative changes: methotrexate is still protonated and we will see cooperativity between the two 'fragments'.

Blundell: In most of the other dehydrogenases isn't the binding of coenzyme before substrate obligatory? You seem to imply that it is not.

Roberts: For dihydrofolate reductase it is not obligatory in the sense that the substrate will bind perfectly well in the absence of coenzyme, although it does bind better in the presence of coenzyme. We do not know whether an ordered mechanism is *kinetically* preferred to a random one. This enzyme presented us with several difficulties in its steady-state kinetics as we found that it exists, in roughly equal amounts, in two conformational forms which interconvert relatively slowly. This phenomenon produces sufficient confusion in the steady-state kinetics to prevent us from establishing with certainty whether kinetically it operates by a random or an ordered mechanism.

References

BRÄNDÉN, C.-I. & EKLUND, H. (1978) Coenzyme-induced conformational changes and substrate binding in liver alcohol dehydrogenase, in *This Volume*, pp. 63-80

MATTHEWS, D. A., ALDEN, R. A., BOLIN, J. T., FREER, S. T., HAMLIN, R., XUONG, N., KRAUT, J., POE, M., WILLIAMS, M. & HOOGSTEEN, K. (1977) Dihydrofolate reductase: X-ray structure of the binary complex with methotrexate. *Science (Wash. D. C.) 197,* 452-455

Polypeptide hormone—receptor interactions: the structure and receptor binding of insulin and glucagon

S. BEDARKAR, T. L. BLUNDELL, S. DOCKERILL, I. J. TICKLE and S. P. WOOD

Laboratory of Molecular Biology, Department of Crystallography, Birkbeck College, University of London

Abstract   Insulin is a small globular protein with a well defined tertiary structure which is closely similar in all species with the exception of certain hystricomorphs such as the guinea pig. Insulin-like growth factor is homologous with insulin and probably has an insulin-like tertiary structure. In contrast glucagon is not a globular protein. It exists as an equilibrium population of conformers with low helix content at physiological concentrations but attains a largely helical conformation on association to trimers.

The receptor binding of insulin depends critically on the correct three-dimensional juxtaposition of groups (A1, A21, B25, etc) and involves both hydrophobic and polar interactions. In insulin-like growth factor part of the insulin receptor region is thought to be buried in extra peptide, so explaining its weak binding to insulin receptors. In contrast the glucagon receptor complex probably involves largely hydrophobic contacts which are possible when a helical conformer is formed.

There are three essential stages in hormone action. First, the hormone must be synthesized correctly and stored in the endocrine tissue; secondly, it must be transported to, and recognize, the 'receptor'; and thirdly it must stimulate a signal leading to a biological response. Many polypeptide hormones have been completely chemically characterized and their biological responses have been widely investigated. Unfortunately, 'receptors' have been less amenable to study.

The most important methodological development in the study of hormone–receptor interactions has been the use of radioactively labelled hormone preparations to identify the receptor (for review see Kahn 1976). Such experiments indicate that polypeptide hormones have receptors which are an integral part of the plasma membrane of the hormone-sensitive cells, in contrast to steroid hormones which must enter the cell. As yet no hormone receptor has been completely purified. The number of receptors is relatively

105

small and about 25 000-fold enrichment is required from a crude membrane fraction. Affinity chromatography, which promised to be useful, has often proved tedious, and many difficulties have been encountered in removing the bound receptor from the affinity column. Even the most highly purified receptor preparation — that of luteinizing hormone (Dufau *et al.* 1975) — is probably only 25% pure and the receptors for insulin (Cuatrecasas 1972) and for glucagon (Giorgio *et al.* 1974) are prepared in extremely small quantities and to lower specific activity.

In view of these difficulties the hormone–receptor complex cannot be easily studied by physical and chemical techniques. An alternative approach is to study the structure of the hormones, their agonists and antagonists, and use these data in combination with receptor binding studies to deduce information about hormone–receptor interactions. In this respect, studies of insulin and glucagon are at an advanced stage. The amino-acid sequences are known and the structures have been studied in the crystal and in solution. In this paper we review the results of structural and the receptor-binding studies on these two pancreatic hormones and on some homologous polypeptides, and discuss models for the hormone–receptor interactions.

INSULIN, PROINSULIN AND INSULIN-LIKE GROWTH FACTOR

Insulin is a polypeptide hormone from the β-cells of the Islets of Langerhans of the pancreas (for reviews see Fritz 1972). It is synthesized as a preprohormone, which contains a hydrophobic NH_2-terminal extension responsible for directing the nascent polypeptide to the endoplasmic reticulum for export (Chan *et al.* 1976). The prohormone is the form in which it is transported to the site of granulation in the Golgi bodies; it is a single chain polypeptide of about 80 amino acid residues (Steiner *et al.* 1969). It is cleaved to give insulin, which comprises two polypeptides, the A-chain of 21 amino acids and the B-chain of 30 amino acids, and stored in granules of zinc insulin.

In response to high circulating concentrations of glucose, amino acids and certain other metabolites, the granules are released into circulation by exocytosis and the insulin finds its way to various tissues including the liver, muscle and fat cells, where it has a generally anabolic effect increasing glucose and amino acid transport into cells and stimulating the synthesis of glycogen and other macromolecules.

Much of the insulin-like activity in serum is not suppressible with guinea pig anti-insulin antibodies. This non-suppressible insulin-like activity (NSILA) has been recently purified and shown to be comprised of a single polypeptide chain (Rinderknecht & Humbel 1976, 1978) with a chemical

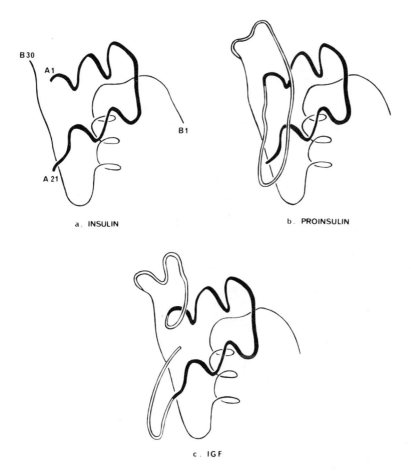

B 30

A 1

B 1

A 21

a. INSULIN

b. PROINSULIN

c. IGF

FIG. 1. Schematic representations of the structure of (a) insulin, based upon the X-ray study of 2-zinc insulin (Blundell *et al.* 1972), (b) proinsulin, based on model-building, and (c) insulin-like growth factor based on model-building (Blundell *et al.* 1978).

structure similar to that of proinsulin. The molecule has strong growth-promoting activity (Megyesi *et al.* 1975) and has been renamed insulin-like growth factor.

X-ray studies of porcine insulin (Blundell *et al.* 1972) have demonstrated that it has the globular structure shown schematically in Fig. 1a and in detail in Fig. 2. The globular structure has two hydrophobic surfaces which are buried during the formation of dimers and 2Zn·insulin hexamers. X-ray studies have demonstrated that most mammalian and fish insulins form similar hexamers, with the exception of hagfish which forms only dimers (Cutfield *et al.* 1974; Peterson *et al.* 1974) and guinea-pig and coypu insulins

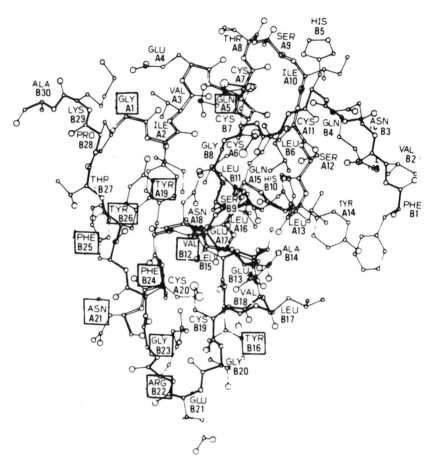

FIG. 2. The structure of insulin based on the X-ray studies of 2-zinc insulin (Blundell *et al.* 1972). The residues with numbers enclosed in boxes are thought to be involved in receptor binding. (Reproduced with permission from Pullen *et al.* 1976).

which form only monomers and probably have a distorted conformation (Wood *et al.* 1975). Circular-dichroic studies indicate that both insulin and proinsulin have a similar three-dimensional structure in solution. The connecting peptide of proinsulin is almost certainly folded over the A-chain on the surface as indicated schematically in Fig. 1b. Model building of the human insulin-like growth factor (Blundell *et al.* 1978) indicated conservation of the insulin fold and hydrophobic core, accommodating a short connecting peptide and an extension at the COOH-terminus of the A-chain, as shown schematically in Fig. 1c.

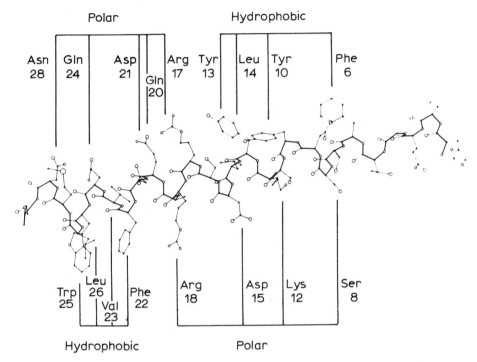

FIG. 3. The structure of glucagon found in cubic crystals (Sasaki *et al.* 1975) indicating the hydrophobic residues involved in formation of trimers and thought to be important in receptor binding. (Reproduced with permission from Blundell *et al.* 1976).

GLUCAGON AND HOMOLOGOUS HORMONES

Glucagon is also a pancreatic hormone synthesized in the α-cells of the Islets of Langerhans (for reviews see Unger 1976). Like insulin it is almost certainly synthesized as a larger prohormone (Tager & Steiner 1973) and is stored in well defined granules. Its biological effects are mainly on the liver, where it leads to a breakdown of glycogen and an increase in circulating glucose concentrations (see Unger 1976).

The hormone is a single polypeptide of 29 amino acids. The conformation of the glucagon molecule in crystals as determined by X-ray analysis (Sasaki *et al.* 1975) is shown in Fig. 3. The molecule is roughly α-helical between residues 6 and 27 and this conformation results in Phe-6, Tyr-10, Tyr-13 and Leu-14 forming one mainly hydrophobic region and Ala-19, Phe-22, Val-23, Trp-25, Leu-26 and Met-27 another. These hydrophobic regions interact to form trimers in the crystals and the trimers are further associated as oligomers of

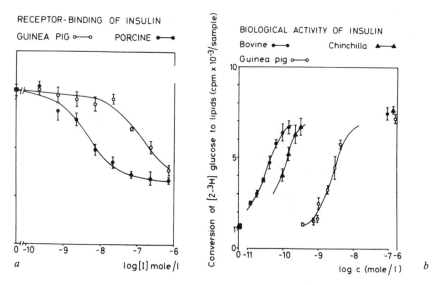

FIG. 4. (*a*) A comparison of the ability of porcine and guinea-pig insulins to inhibit the binding of
[125]I-labelled insulin to insulin receptors of fat-cell membranes: (*b*) a demonstration of the stimulation by bovine, chinchilla and guinea-pig insulins of the incorporation of [3H]glucose into lipids
by rat fat cells shown as a logarithmic function of insulin concentration in incubation, from the
work of S. Gammeltoft and J. Gliemann. (Reproduced with permission from Pullen *et al.* 1975).

cubic symmetry. Residues 1–5 are not constrained by intermolecular interactions and appear to be flexible. Optical rotatory dispersion and circular
dichroism (Panijpan & Gratzer 1974) indicate that glucagon has little secondary structure in aqueous solution at high dilutions characteristic of those in
circulation. It probably exists as an equilibrium population of conformers.
The percentage of the helical conformers must be low but it is increased by
self-association to trimers (Blanchard & King 1966) and in non-aqueous
solvents (Bornet & Edelhoch 1971).

Secretin and vasoactive intestinal polypeptide (VIP) are homologous in
sequence with glucagon (Mutt *et al.* 1970; Mutt & Said 1974). The general
distribution of hydrophobic and hydrophilic residues in all three hormones is
similar but the nature of the individual side-chains differs. If secretin and VIP
were to form helical conformers they would also have two hydrophobic
regions similar to those of glucagon (Blundell *et al.* 1976).

INSULIN AND GLUCAGON RECEPTORS

Studies with labelled hormone preparations indicate that the receptors are
on the plasma membrane. Fig. 4a compares the ability of porcine and

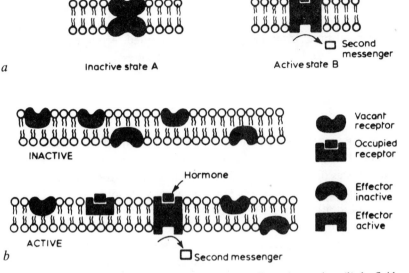

a   Inactive state A   Active state B

INACTIVE

Hormone

ACTIVE

b   Second messenger

Vacant receptor

Occupied receptor

Effector inactive

Effector active

FIG. 5. (*a*) The two-state model for hormone–receptor–effector interaction; (*b*) the fluid-mosaic-membrane model of the hormone–receptor–effector interactions. (Reproduced with permission from Blundell & Wood 1976).

guinea-pig insulins to inhibit the binding of ^{125}I-labelled insulin to insulin receptors of the membranes of fat cells and Fig. 4b demonstrates the stimulation by bovine, chinchilla and guinea-pig insulins of the incorporation of [^{3}H]glucose into lipids by rat fat cells shown as a logarithmic function of insulin concentration in the incubation. Experiments of this kind demonstrate that there is a finite number of receptors available to the hormone but that only a small percentage of these need be occupied for a full biological response.

Attempts to extract the receptor molecules (Cuatrecasas 1972; Giorgio *et al.* 1974) indicate that the receptors are probably integral membrane proteins. The second messenger for glucagon action in liver cells is cyclic AMP and thus the effector or catalytic molecule is identified as the enzyme adenylate cyclase (for reviews see Unger 1976) situated on the inner side of the cell membrane. For insulin the nature of the second messenger is still in dispute.

The details of the receptor–effector interaction for glucagon have been studied extensively and several models have been presented which fit the experimental data. Two of the best models are depicted simply in Fig. 5; in reality they are more complex, especially with regard to the involvement of guanine nucleotides in the linkage. The simple two-state equilibrium model (Fig. 5a) has been superseded by the fluid-membrane model (Fig. 5b) in which the receptors are free to move in the plane of the extracellular side of the cell membrane. The effector molecule is free to move on the intracellular side of the membrane. Interactions between occupied receptor and the adenylate cyclase lead to the production of the second messenger.

HORMONE–RECEPTOR INTERACTIONS

We have shown elsewhere (Blundell *et al.* 1972; Blundell & Wood 1975; Sasaki *et al.* 1975) that insulin and glucagon are probably stored in the pancreatic cells as oligomers packed in crystalline or amorphous granules. In circulation, insulin and glucagon granules slowly dissolve and at concentrations of about 10^{-10} mol/l the oligomeric forms (hexamers and trimers) must dissociate to give monomers. This makes the monomer the most likely candidate for the active species, but it is still possible that it reassociates at or in the proximity of the receptor. However, reassociation seems to be an unlikely requirement as several insulins which show little ability to dimerize (for example, guinea-pig insulin and some nitrated insulins) are biologically active, albeit weakly so (Blundell *et al.* 1972).

The biological activity of insulin is usually directly proportional to the receptor binding, as measured by the ability of the insulin to displace radioactively labelled insulin from the receptor. The activity depends on the integrity of the three-dimensional structure of insulin; no insulin with a disturbed tertiary structure is fully active (Blundell *et al.* 1972; Pullen *et al.* 1976). Some residues involved in dimerization, in particular Phe-B24 and Phe-B25, and possibly the adjacent B12, B16 and B26, probably bind the receptor initially through non-covalent interactions (see Fig. 2). A1, A19 and A21 have also been postulated to bind the receptor; however, these residues can be modified without complete loss of activity and evidence is accumulating that they are on the periphery of the receptor-binding region. Secondary conformational changes and degradation of insulin as a necessity for the biological activity remain possibilities. The importance of the tertiary structure is clearly a consequence of the wide separation on the sequence of groups important to receptor binding which must be brought together in the correct relative positions.

In the insulin–receptor complex, it is probable that the receptor has a concave surface complementary to that of the hormone. The surface may have maximum dimensions of 20×20 Å but as residues on the periphery of the insulin-receptor-binding region are not essential the interaction at this point may not be close. The receptor almost certainly must have a large hydrophobic region complementary to that formed by residues Phe-B24, Phe-B25, Tyr-B26, Val-B12 and possibly part of Tyr-A19. It must also be able to satisfy the hydrogen-bond donors of the main chain of residues B24–B26 and so may have a polypeptide chain capable of forming an antiparallel β-sheet interaction. On the periphery it must contain a series of charged groups which can form ionic interactions with the A1 α-amino, A4 glutamyl carboxylate, A21 α-carboxylate, B22 guanidinium, and B13 glutamyl carboxylate groups. It thus involves a larger surface area and more extensive interactions than exhibited in the dimer which account for the higher association constant.

It is likely that the homologous polypeptides have similar receptor interactions. Insulin-like growth factor will bind to insulin receptors but with reduced affinity; on the other hand receptors for insulin-like growth factor bind that hormone more strongly than insulin (Megyesi et al. 1975). This cross-reactivity probably depends on the common tertiary structure and hydrophobic residues whereas the specificity may be mediated by the amino-acid residues in the connecting peptide and the COOH-terminus of the A-chain of the growth factor. In a complementary way the receptor for the growth factor must resemble that of insulin in some of the hydrophobic regions and the hydrogen-bonding capacity but must be different in the regions complementary to the charged groups such as A1 α-amino and A21 α-carboxylate, and to the Phe-B25 which becomes Tyr.

Studies on glucagon show that almost the whole molecule is required for full biological activity although loss of the NH_2-terminal histidine has less effect on the receptor binding than on the potency (Rodbell et al. 1971). It has been suggested that receptor binding involves hydrophobic interactions and that a helical conformer is induced or stabilized at the receptor (Sasaki et al. 1975). This conformer has organized hydrophobic regions which might interact with complementary regions of the receptor. Other charged interactions, that is with Arg-17 and Arg-18, may further increase the receptor binding, but interaction of the NH_2-terminal residues could enhance the biological response without a proportional increase in the stability of the hormone–receptor complex. The ease of forming a helical conformer in the correct environment means that a stable tertiary structure is not required in glucagon.

The specificity of glucagon and secretin for their own receptors may be a result of the difference of the hydrophobic residues — mainly aromatic in glucagon but more aliphatic in secretin — although the conformer at the receptor in each case may be similar (Blundell et al. 1976. Thus insulin and glucagon have two contrasting mechanisms for recognizing the receptor: one requires a relatively rigid conformation and the other depends on the induction of the conformer.

However, both involve residues which are buried in the oligomeric forms and give rise to a stable storage form.

ACKNOWLEDGEMENTS

We are grateful to Professor Dorothy Hodgkin, Dr Guy Dodson, Dr Dan Mercola, Dr Martin Rodbell, Professor Rene Humbel, Dr Jorgen Gliemann, Mr John Jenkins, Mr Trevor Sewell and many other colleagues and friends for useful discussions about the ideas presented in this paper. We thank Mrs Margaret Crowe, Mrs Liz Wood and Miss Sue Kennett for technical assistance. The Science Research Council (UK) supported this work.

References

BLANCHARD, M. H. & KING, M. V. (1966) Evidence of association of glucagon from optical rotatory dispersion and concentration-difference spectra. *Biochem. Biophys. Res. Commun.* 25, 298-303

BLUNDELL, T. L. & WOOD, S. P. (1975) Is the evolution of insulin Darwinian or due to selectively neutral mutation? *Nature (Lond.)* 257, 198-203

BLUNDELL, T. L. & WOOD, S. P. (1976) Membrane receptors. *New Sci.,* 670-671

BLUNDELL, T. L., DODSON, G. G., HODGKIN, D. C. & MERCOLA, D. A. (1972) Insulin: the structure in the crystal and its reflection in chemistry and biology. *Adv. Protein Chem.* 26, 279-402

BLUNDELL, T. L., DOCKERILL, S., SASAKI, K., TICKLE, I. J. & WOOD, S. P. (1976) The relation of structure to storage and receptor binding of glucagon. *Metabolism* 25, 1331-1336

BLUNDELL, T. L., BEDARKAR, S., RINDERKNECHT, E., & HUMBEL, R. E. (1978) Insulin-like growth factor: a model for tertiary structure accounting for immunoreactivity and receptor binding. *Proc. Natl. Acad. Sci U.S.A.* 75, 180-184

BORNET, H. & EDELHOCH, H. (1971) Polypeptide hormone interaction: glucagon detergent interaction. *J. Biol. Chem.* 246, 1785-1792

CHAN, S. J., KEIM, P. & STEINER, D. F. (1976) Cell-free synthesis of rat preproinsulins; characterization and partial amino acid sequence determination. *Proc. Natl. Acad. Sci. U.S.A.* 73, 1964-1968

CUATRECASAS, P. (1972) Affinity chromatography and purification of the insulin receptor of liver cell membranes. *Proc. Natl. Acad. Sci. U.S.A.* 69, 1277-1280

CUTFIELD, F. J., CUTFIELD, S. M., DODSON, E. J., DODSON, G. G. & SABESAN, M. N. (1974) Low resolution crystal structure of hagfish insulin. *J. Mol. Biol.* 87, 23-30

DUFAU, M. L. RYAN, D. W., BAUKAL, A. J. & CATT, K. J. (1975) Gonadotropin receptors. Solubilization and purification by affinity chromatography. *J. Biol. Chem.* 250, 4822-4824

FRITZ, I. (1972) *Insulin Action,* Academic Press, New York and London

GIORGIO, N. A., JOHNSON, C. B. & BLECHER, M. (1974) Hormone receptors. 3. Properties of glucagon-binding protein isolated from liver plasma membranes. *J. Biol. Chem.* 249, 428-437

KAHN, C. R. (1976) Membrane receptors for hormones and neurotransmitters. *J. Cell Biol.* 70, 261-286

MEGYESI, K., KAHN, C. R., ROTH, J., NEVILLE, D. M., NISSLEY, S. P., HUMBEL, R. E. & FROESCH, E. R. (1975) The NSILA-s receptor in liver plasma membranes. Characterization and comparison with the insulin receptor. *J. Biol. Chem.* 250, 8990-8996

MUTT, V. & SAID, S. I. (1974) Structure of the porcine vasoactive intestinal octacosapeptide. The amino acid sequence. Use of kallikrein in its determination. *Eur. J. Biochem.* 42, 581-589

MUTT, V., JORPES, J. E. & MAGNUSSON, S. (1970) Structure of porcine secretin. The amino acid sequence. *Eur. J. Biochem.* 15, 513-519

PANIJPAN, B. & GRATZER, W. B. (1974) Conformational nature of monomeric glucagon. *Eur. J. Biochem. 45*, 547-554

PETERSON, J. D., COULTER, C. L., STEINER, D. F., EMDIN, S. O. & FALKMER, S. (1974) Structural and crystallographic observations on hagfish insulin. *Nature (Lond.) 251*, 239-240

PULLEN, R. A., JENKINS, J. A., TICKLE, I. J., WOOD, S. P. & BLUNDELL, T. L. (1975) The relation of polypeptide hormone structure and flexibility to receptor binding: the relevance of X-ray studies on insulins, glucagon, and human placental lactogen. *Mol. Cell. Biochem. 8*, 5-20

PULLEN, R. A., LINDSAY, D. G., WOOD, S. P., TICKLE, I. J., BLUNDELL, T. L., WOLLMER, A., KRAIL, G., BRANDENBURG, D., ZAHN, H., GLIEMANN, J. & GAMMELTOFT, S. (1976) Receptor-binding region of insulin. *Nature (Lond.) 259*, 369-373

RINDERKNECHT, E. & HUMBEL, R. E. (1976) Amino-terminal sequences of two polypeptides from human serum with nonsuppressible insulin-like and cell-growth-promoting activities: evidence for structural homologies with insulin B chain. *Proc. Natl. Acad. Sci. U.S.A. 73*, 4379-4381

RINDERKNECHT, E. & HUMBEL, R. E. (1978) *J. Biol. Chem.*, in press

RODBELL, M., BIRNBAUMER, L., POHL, S. L. & SUNDBY, F. (1971) The reaction of glucagon with its receptor: evidence for discrete regions of activity and binding in the glucagon molecule. *Proc. Natl. Acad. Sci. U.S.A. 68*, 909-913

SASAKI, K., DOCKERILL, S., ADAMIAK, D. A., TICKLE, I. L. & BLUNDELL, T. L. (1975) X-ray analysis of glucagon and its relationship to receptor binding. *Nature (Lond.) 257*, 751-757

STEINER, D. F., CLARK, J. L., NOLAN, C., RUBENSTEIN, A. H., MARGOLIASH, E., ATNE, B. & OYER, P. E. (1969) Proinsulin and the biosynthesis of insulin. *Rec. Progr. Horm. Res. 25*, 207

TAGER, H. S. & STEINER, D. F. (1973) Primary structure of the proinsulin connecting peptides of the rat and the horse. *Proc. Natl. Acad. Sci. U.S.A. 70*, 2321-2325

UNGER, R. (ed.) (1976) Glucagon Symposium. *Metabolism 25 (Suppl. 1)*, 1303-1533

WOOD, S. P., BLUNDELL, T. L., WOLLMER, A., LAZARUS, N. R. & NEVILLE, R. W. J. (1975) The relation of conformation and association of insulin to receptor binding; x-ray and circular-dichroism studies on bovine and hystricomorph insulins. *Eur. J. Biochem. 55*, 531-542

Discussion

Blow: Would you explain how you assign the surface of the insulin monomer that you think binds to the receptor? What is the evidence that the insulin monomer maintains the conformation that is found in the crystalline hexamer?

Blundell: As insulins from different organisms are not species specific, there must be a common feature to the different insulins. The invariant regions of the insulin molecule comprise the disulphide bridges and the hydrophobic core, which are important to the tertiary structure, and in addition a surface area of the molecule, part of which is involved in dimerization. If all insulins bind to each other's receptors in the same way, it is probable that this surface region binds to the receptor, if the monomer is the active form. The evidence for the binding of monomers by receptors is that guinea-pig insulin, tetra-nitroinsulin and several other modified insulins do not self-associate but are nevertheless biologically active. When insulins are specifically modified at various functional groups close to this putative receptor-binding region, for instance by addition of groups of different size and different charge at A1, we find that no insulin with a conformation unlike the native one is fully biologically active. All insulins that have decreased activity either have a changed conformation or have groups added which directly interfere with the receptor-binding region. We have studied each modified insulin by several different receptor assays, by spectroscopic techniques for solutions, and by crystallization and determination of the three-dimensional structure by X-ray crystallography. In this way we have tried to map the surface of the insulin molecule important for receptor binding.

Richards: The only way you will really know is when you know the structure of the receptor. A standard difficulty is that receptors are present in relatively small amounts and of those present only a minute fraction are likely to be of any functional importance. This seems to be generally true whether the receptors are of bacterial or mammalian origin. How can we get round this problem? Obviously we shall not be able to isolate the receptors for some time yet.

Blundell: For insulin, most cells have several thousand receptors per cell although the number varies from one cell to another. The cells which are most amenable to extraction of the receptors, such as turkey erythrocytes, do not have a well defined biological response. Such cells seem to have more receptors than they need but that does not mean to say that not all the receptors are competent to bind the hormone. The surplus of receptors — the spare receptors — increases the probability of the hormone's binding a receptor and compensates for the fact that the receptor-binding affinities (about 10^{-9} mol/l) are lower than the circulating concentrations of the hormone (which are about 10^{-10}–10^{-11} mol/l).

With regard to extraction of receptors, in particular the insulin receptor, it has proved difficult to repeat many of the reported extractions, especially with respect to removal of the receptor from the affinity column. Also in the isolation procedure one isolates only the regulator and not the effector enzyme such as adenylate cyclase. The assays measure only the binding of the hormone. Therefore, one may isolate the hormone-degrading enzyme and there are some indications that this may have happened with insulin. In the absence of any acceptable preparation of insulin or glucagon receptors the experiments I have reported become extremely important.

Dunitz: Would it be too much of an exaggeration to say that these receptors are at present concepts rather than entities?

Blundell: Hormones do have a biological response, so there must be some interaction.

Dunitz: Yes, and that is why you need the concept of a receptor.

Blundell: One of the best examples of a purified receptor is that for luteinizing hormone; that is probably the most pure (about 25%) receptor that has been extracted — but only in microgram quantities. Receptors are not as mythical as might be inferred from your question.

Williams: Insulin can be crystallized with two, four or six zinc atoms. We know from the n.m.r. evidence as well as from the crystal structures that the end of the B-chain of insulin is mobile (Williamson *et al.* 1978).

Blundell: I agree, but several residues of the NH_2-terminus of that chain may be deleted without affecting the biological potency.

Williams: Did you derive the structure of the monomer from the crystal structure of the hexamer?

Blundell: Yes. We have several pieces of evidence that the structure in the two-zinc hexamer is relevant to the monomer. First, crystallographic evidence for the molecules of hagfish insulin and porcine insulin in cubic crystals — where hexamers are not formed — shows that the tertiary structure is similar to that of one of the insulin molecules in the two-zinc hexamers, although the flexible NH_2-terminus of the B-chain is in a different position. Secondly, we can consider spectroscopic techniques to determine the structure in solution. The far-ultraviolet circular dichroic spectrum shows a small change from monomer to dimer to hexamer but this can be accounted for by small changes perhaps in the environment or accessibility of the solvent to the helix and by movement of the flexible NH_2-terminus of the B-chain. The near-ultraviolet circular dichroic spectrum changes as predicted for the changes in the environments of the tyrosyl and phenylalanyl groups. Theoretical calculations on the near-ultraviolet circular dichroism (Strickland & Mercola 1976) show that the changes in that spectrum can be explained if one assumes that the monomer has the same tertiary structure as the hexamer and that only the environment changes. Note that the structure of the four-zinc insulin hexamers occurs at high chloride concentrations and changes in the circular dichroism are also observed in these conditions. Thus, the conformation changes at high protein concentrations and at high ionic strengths but there is no evidence that this is relevant to the physiological situation which is characterized by low insulin concentrations and low ionic strength.

Studies on structure and function indicate that the 'central' part of insulin must have a native-like conformation to bind to the receptor. However, the conformation may still change once the hormone–receptor complex is formed, and this may be necessary for biological response. At present our studies on receptor binding cannot rule out this eventuality. However, the most flexible parts of the insulin molecule, such as the B-chain terminal residues, can be removed with little change in potency. The flexible A-chain loop

(A8–A10) can also be modified with no change in activity. Flexible parts of the structure seem not to be important for activity.

Williams: The release and action of insulin in the human body are relatively slow.

Blundell: The half-life of insulin and glucagon is about 3–6 min.

Williams: In view of the discussion about the packing of insulin in crystals *in vivo*, it is instructive to compare the packing of a small hormone, adrenalin (Daniels *et al.* 1974). The reason that insulin is stored as crystals is that a crystal contains a large amount of material without it being osmotically active. Surfaces of molecules are thus useful for reasons that have nothing to do with receptors — namely, self-crystallization. However, crystallization allows only *slow* release. A small hormone like adrenalin is needed rapidly. Its storage brings us back to the discussion of the protein, chromogranin, which is a random coil and which does not crystallize. Packing small molecules (e.g. adrenalin) with random-coil proteins (such as chromogranin) makes a loose gel. Consequently the concentration of adrenalin in solution is high, but the hormone is not effectively active osmotically because the protein network encases it. This 'packing' lowers the osmotic pressure in a totally different way from that used in insulin storage.

Kenner: Some of the questions raised seem to be based on the idea that there was something inherently improbable about only a portion of such a large molecule being necessary. That is not true; in some other peptide hormones only a small portion of the molecule is necessary for biological activity.

Blundell: Several groups, including yours, Professor Kenner, are synthesizing non-peptide analogues of the section of the molecule that we have suggested binds to the receptor. Such an insulin analogue might be administered orally (a major problem for diabetics, of course, is that insulin itself cannot be taken orally). So the use of such models may have practical use far beyond the theoretical value of receptor-binding studies.

Wüthrich: N.m.r. studies of glucagon in dilute aqueous solution provide direct evidence that the hydrophobic intermolecular interactions in the crystal may be essential for the stabilization of the helical structure. We have studied glucagon in 0.1 mmol/l aqueous solution at pH 10.5 by high-resolution n.m.r. spectroscopy. The fact that only one conformation was observed in the n.m.r. spectra indicates that the solution conformation corresponds to one manifold of rapidly interconverting conformations (Wüthrich 1976). The n.m.r. data also showed that monomeric glucagon in aqueous solution is predominantly in an extended flexible form (A. Bundi, Ch. Boesh, M. Oppliger, R. Andreatta & K. Wüthrich, unpublished results). However, the spectrum also revealed evidence of local rigidity of structure. We were fortunate in being able to locate the portion of the structure that was not extended and flexible as being the sequence Phe-Val-Glu-Trp-Leu (residues 22–26). The local spatial structure in this peptide fragment was characterized (Wüthrich 1976) on the basis of the ring-current shifts experienced by individual protons due to proximity to the indole ring of Trp-25, the coupling constants $^3J_{\alpha\beta}$ for both the tryptophyl and valyl side-chains, and the long-range spin–spin coupling between the β-methylene protons and indole-ring protons of tryptophan, which were observed in the two-dimensional ^1H n.m.r. spectrum of this peptide (Nagayama et al. 1977).

The spatial solution structure for the peptide fragment 22–26 in glucagon which we thus characterized (and which will be described in detail elsewhere; A. Bundi, Ch. Boesch, M. Oppliger, R. Andreatta & K. Wüthrich, as yet unpublished results) is not compatible with the assumption that the α-helix observed in the crystal structure is maintained in solution.

In an α-helix the side-chains of next-nearest residues form an angle of about 180° and the side-chain of valine-23 could thus not be located sufficiently close to the side-chain of tryptophan-25 to experience a sizable high-field ring-current shift.

As a general comment I should add that I do not believe that we have to assume *a priori* that the polypeptide hormone has to contain a lipophilic surface to be bound by a membrane receptor. Therefore, the results on the solution conformation might serve just as well as the crystal data as the basis for hypotheses about possible conformations of glucagon bound to the receptor site. To investigate further this question we are at present studying glucagon in deuteriated detergent micelles, which give well resolved ^1H n.m.r. spectra of the micelle-bound polypeptide chain (Ch. Boesch, L. R. Brown & K. Wüthrich, unpublished results, 1977).

Williams: Is there any activity associated with the pentapeptide?

Wüthrich: We have not yet obtained an unambiguous answer to this question. Originally we started to work on the synthetic human parathyroid hormone fragment 1–34 (Bundi et al. 1976). The n.m.r. study on this hormone led to similar conclusions on the spatial solution structure to those I have just described for glucagon. In human parathyroid hormone, the sequence Arg-Val-Glu-Trp-Leu (residues 20–24) corresponds to the fragment 22–26 of glucagon. When we investigated the glycolytic activity of various partial sequences of human parathyroid hormone, we found evidence that all fragments which contained the pentapeptide 20–24 might retain some activity. Since glucagon contained a closely-similar pentapeptide fragment, we turned our attention also to this molecule.

Blundell: The fragment containing the last 10 residues of the glucagon molecule, which includes the pentapeptide, has less than 0.1% of biological response. We tried to characterize the glucagon-receptor-binding process. First, the evidence suggests that it is driven by entropy and that it has a dependence on urea characteristic of a hydrophobic interaction. Secondly, active analogues can form helices and those that cannot form helices are inactive, as suggested by our model.

Williams: A parallel finding to this is the discovery of enkephalin, an active pentapeptide. N.m.r. spectroscopy of that fragment in aqueous solution shows that it has no really fixed conformation (B. A. Levine & R. J. P. Williams, unpublished results). Small residual elements of structure only can be seen. The biological activity increases greatly when the peptide is lengthened to 29 residues (i.e. to β-endorphin): increasing the chain length from 5 to 27 or 28 residues hardly raises the activity but on addition at the 29th residue it really increases. The reason for that change is the introduction of a positively charged amino acid (number 29). Here it is not a hydrophobic effect that is important. (This work is due to Dr D. G. Smyth of Mill Hill, London).

Blundell: I am not generalizing with respect to the hydrophobic effect nor about the relevance of the data on crystal structure to biological activity. For enkephalins our (unpublished) crystal data indicate a β-sheet conformation which cannot be attained by all active analogues. However, I should caution comparison of enkephalins and β-endorphin because the biological roles and responses differ. The enkephalins are probably neurotransmitters. β-Endorphin is degraded too slowly to be a neurotransmitter.

Cornforth: If the mechanism of the effect of binding of a hormone to a receptor is a conformational change in the receptor, isn't that easier to understand for a conformationally-stable molecule like insulin than for a flexible one like glucagon?

Blundell: That is not necessarily correct. For instance, RNAase-S is a good model for a flexible hormone–receptor complex. The S-peptide is equivalent to, say, the glucagon molecules and the rest of the conformation of the polypeptide chain is equivalent to the receptor. As the S-peptide binds, it undergoes a large-conformational change but is effective in defining the conformation in the rest of the enzyme.

References

BUNDI, A., ANDREATTA, R., RITTEL, W. & WÜTHRICH, K. (1976) Conformational studies of the synthetic fragment 1–34 of human parathyroid hormone by NMR techniques. *FEBS (Fed. Eur. Biochem. Soc.) Lett. 64*, 126-129

DANIELS, A., KORDA, A., TRANSWELL, B., WILLIAMS, A. & WILLIAMS, R. J. P. (1974) The internal structure of the chromaffin granule. *Proc. R. Soc. Lond. B Biol. Sci. 187*, 353-361

NAGAYAMA, K., WÜTHRICH, K., BACHMANN, P. & ERNST, R. R. (1977) Two-dimensional J-resolved ¹H n.m.r. spectroscopy for studies of biological macromolecules. *Biochem. Biophys. Res. Commun. 78*, 99-105

STRICKLAND, E. H. & MERCOLA, D. (1976) Near-ultraviolet tyrosyl circular dichroism of pig insulin monomers, dimers, and hexamers. Dipole–dipole coupling calculations in the monopole approximation. *Biochemistry 15*, 3875-3884

WILLIAMSON, A. R., BENTLEY, G. & WILLIAMS, R. J. P. (1978). *Biochemistry,* in press

WÜTHRICH, K. (1976) *NMR in Biological Research: Peptides and Proteins*, North-Holland, Amsterdam

Perturbations of model protein systems as a basis for the central and peripheral mechanisms of general anaesthesia

M. J. HALSEY, F. F. BROWN* and R. E. RICHARDS*

*Division of Anaesthesia, Clinical Research Centre, Harrow, Middlesex and *Biochemistry Department, University of Oxford, Oxford*

Abstract   Protein perturbations associated with anaesthetic interactions are relevant to: (*a*) the central molecular mechanisms of general anaesthesia; (*b*) the molecular basis of physiological selectivity and anaesthetic specificity of the many 'side-effects' of anaesthesia; (*c*) the use of anaesthetic agents as selective hydrophobic probes for the study of protein structures and activities in detail.

Small but specific protein perturbations have been studied with various nuclear magnetic resonance procedures with haemoglobin as a model protein to establish the 'ground-rules' for anaesthetic–macromolecule interactions. The correlation of one aspect of these perturbations with anaesthetic potency and hydrophobic solubility indicates that hydrophobic pockets in proteins can behave like bulk-lipid phases in terms of their solubility characteristics. Other aspects appear to depend on physical characteristics such as size, geometry, structure and composition of the individual agents. These data support the hypothesis that anaesthetic actions can be explained on a molecular basis by direct interactions with proteins in addition to lipid and aqueous effects.

Protein conformational perturbations produced by clinical concentrations of general anaesthetics are potentially relevant to three areas of investigation: first, the central molecular mechanisms of general anaesthesia which occur in the synaptic regions of the neuronal networks; second, the peripheral mechanisms of anaesthesia associated with the 'side-effects' and toxicity of specific agents; third, the use of anaesthetic agents as selective hydrophobic probes for the study of protein structures and activities in detail.

The mechanism of general anaesthesia at a molecular level is still far from clear. It may be ascribed either to changes in lipid thermotropic behaviour or to a malfunction of neuronal proteins due to direct anaesthetic interactions, or to a combination of the two. The inhalational anaesthetics are usually con-sidered to be the archetype of non-specific drugs: that is, their central actions do not depend on their precise chemical structures interacting with an array of spatially organized receptor sites. However, this traditional view may have to be modified to accommodate recent evidence on the selectivity and specificity of anaesthetic 'side-effects' which are superimposed on the central mechanisms of anaesthesia (Halsey 1977). This selectivity of action can be explained on a molecular level by the diversity of protein structures associated with different functions.

The general anaesthetic potencies *in vivo* of all the inhalational agents correlate better with their lipophilic solubilities than with any other physical property so far studied. This has led to the assumption that anaesthesia occurs when a critical molar, or volume, concentration is attained by hydrophobic sites of action. Thus, anaesthetic potency is thought to depend on either the concentration of the agent (mole fraction) at the site of action or the product of this concentration and the molar volume of the anaesthetic (volume fraction). The most probable solubility characteristics of the sites of action have been quantitated in terms of solubility parameters with values of 18.4 ± 2 and of 20.5 ± 2 $(J/cm^3)^{1/2}$ in the case of the critical molar and volume concentration assumptions, respectively (Miller *et al.* 1972).

These characteristics exclude a significant proportion of the anaesthetic site of action having polar characteristics and imply that anaesthetics act at hydrophobic sites in lipid bilayers or macromolecules.

Another feature of general anaesthesia is that it can be 'antagonized' by application of high pressures of the order of 100 atm. Estimates of the compressibility of the site of action (6×10^{-5} atm^{-1}) agree with those predicted from the solubility characteristics (Miller *et al.* 1973). However, quantitation of the pressure–anaesthetic interactions in rats and mice has revealed details of this phenomenon which are not consistent with the traditional unitary view that all anaesthetics act at the same type of molecular site (M. J. Halsey, B. Wardley-Smith & C. J. Green, unpublished work). It may be that future hypotheses for molecular mechanisms of anaesthesia will have to include multi-sites of action.

ANAESTHETIC EFFECTS ON PROTEIN FUNCTION

As long ago as 1944, Östergren postulated that lipid-soluble agents could affect 'protein chain folding' and that this was related to 'narcosis'. There is now considerable evidence that the inhalational agents can interact with many proteins and produce functional changes. The effects of the environment and inhibitors on a range of different proteins give some clues to the relative importance of anaesthetic perturbations of secondary, tertiary and quaternary structures.

Anaesthetic depression of enzyme activity has been investigated in several cases. For example, glutamate dehydrogenase is reversibly depressed by halothane ($CF_3 \cdot CHBrCl$) and methoxyflurane ($MeO \cdot CF_2 \cdot CHCl_2$) whereas many of the enzymes in the Embden–Meyerhoff pathway are relatively un-affected (Brammall et al. 1974). One enzyme system that has been extensively studied is the luminescent reaction in luminous bacteria, which is activated by low concentrations and reversibly depressed by clinical concentrations of inhalational anaesthetics (White & Dundas 1970; Halsey & Smith 1970). Detailed in vitro biochemistry with 'lipid-free' luciferase has identified the probable site of anaesthetic action (Middleton 1973; White et al. 1975). The structure of this enzyme has not been elucidated fully but includes a highly hydrophobic active centre with an associated reactive thiol group (Nicoli et al. 1974). Such a structure in the active centre leads to the working hypothesis that protein functions are sensitive to anaesthetics if, and only if, there are suitable hydrophobic areas available at or near the active centres. White and his colleagues reviewed the structure of those enzymes whose secondary and tertiary structures are known, and selected papain and bromelain as examples of proteins with hydrophobic active centres containing active thiol groups, in contrast with the properties of lysozyme and α-chymotrypsin. The functions of papain and bromelain are reversibly depressed by clinical concentrations of anaesthetics whereas the functions of the 'non-hydrophobic' active-centre proteins are unaffected (King et al. 1977).

The relationships between different anaesthetic-induced functional changes in the same protein are not always clear. Thus, the oxygen-binding function of haemoglobin does not appear to be affected by clinical inha-lational anaesthetics (Millar et al. 1971; Weiskopf et al. 1971) even at the extremes of the physiological pH range (M. J. Halsey & B. Minty, unpub-lished work). However, high partial pressures of nitrogen appear to increase the oxygen affinity of haemoglobin (Kiesow 1974) but low partial pressures of dichloromethane affect its capacity to bind carbon monoxide (Settle 1975).

ANAESTHETIC EFFECTS ON PROTEIN CONFORMATION

Relatively few direct investigations have been made of the underlying conformational perturbations produced by anaesthetics. X-ray crystallo-graphy has been used to locate the binding site of xenon in haemoglobin and of both xenon and cyclopropane in myoglobin (Schoenborn 1965, 1968). Optical rotatory dispersion has been used to demonstrate the tertiary struc-tural changes produced by high concentrations of anaesthetics on bovine serum albumin and β-lactoglobulin (Balasubramanian & Wetlaufer 1966).

Similar studies have also been done on various forms of haemoglobin — including the isolated β-chains (Laasberg & Hedley-White 1971) — and on myosin, aldolase, human serum albumin, ribonuclease and lysozyme (Leuwenkroon-Strosberg 1973). The helical content of the haemoglobins, aldolase and myosin, appeared to be decreased by 3–10% on exposure to all anaesthetics used but significant effects on the secondary structure of the other proteins could not be detected.

NUCLEAR MAGNETIC RESONANCE SPECTROSCOPY

This paper is devoted to aspects of anaesthetic perturbations of the conformation of a model protein system. To observe perturbations of this kind we need a technique that is sensitive to small structural changes in a protein which, if possible, is in its physiological state: i.e. membrane-bound proteins should be studied bound to membranes, and soluble proteins should be studied in solution. Departure from these existing conditions may alter or restrict the access of inert anaesthetic molecules which do not bind in the accepted sense but merely undergo selective 'solution' within the protein interior.

At present it is not possible to make direct detailed studies on membrane-bound proteins and there are limitations in assessing potential perturbations in membrane-bound proteins from the behaviour of soluble proteins. However, even though part of the surface of a membrane-bound protein may be covered by hydrophobic groups, the underlying concept of protein distortion by invasion of hydrophobic areas beneath the surface is probably still a valid one, although the effects of such distortions are less easily assessed as they may affect not only the conformation of the active site but the mobility of the protein within the lipid bilayer.

One of the techniques most sensitive to small structural perturbations in soluble proteins is spin-echo Fourier-Transform nuclear magnetic resonance spectroscopy (Campbell *et al.* 1975). If this technique is to be used effectively, the protein must produce a well-resolved spectrum containing at least a few lines capable of unique assignment so that the properties observed are those of one particular nucleus, even if its actual assignment is unknown. This imposes an upper limit on the molecular weight of the protein of about 70 000. At the same time it is an advantage to have a large protein with various hydrophobic pockets so as to provide an opportunity for study of site preferences determined by the size, shape and composition of the anaesthetic molecule. Furthermore, allowance must be made for the basic insensitivity of n.m.r. spectroscopy in terms of signal acquisition, so that the protein must be soluble (e.g. > 1 mmol/l) and available in reasonably large quantities (e.g. 50 mg of a high-molecular-weight protein per sample).

A protein that fits all these requirements is haemoglobin. It has a further advantage that, being composed of subunits whose relative movement is germane to its function, it offers further scope for comparison with, for example, the acetylcholine receptor protein, whose function may also be associated with the movement of subunits.

The application of spin-echo techniques to Fourier-Transform n.m.r. spectroscopy greatly enhances the basic sensitivity of the method to small structural changes. Not only is the signal frequency sensitive to the precise magnetic environment of a nucleus within the protein framework, but the length of time taken by a particular nucleus to dissipate its spin energy either individually to another similar nucleus (T_2) or collectively to the lattice (T_1) will critically control the line width and line intensity in a given set of experimental conditions. It is these latter, more delicate, parameters to which the spin-echo experiment is particularly sensitive through its two delay variables, the interpulse spacing (sensitive to T_2) and the scan repetition rate (sensitive to T_1). These parameters will be most affected by the subtle changes in conformation and mobility caused by anaesthetic perturbations. Although such changes in the relaxation times, and therefore in the line intensity, will be small compared with the normal peak height of a perturbed resonance, it is possible, by use of fairly long delay times between the 90° and 180° pulses, to amplify these effects greatly so that these small intensity changes now become a significant fraction of the intensity observed in these conditions.

When anaesthetics interact with haemoglobin a series of perturbations occur but the ones which will be considered here are those from the C-2 protons from histidyl groups on the surface of the molecule. The reason for this is not that their perturbations are more pronounced but because their resonances are the only ones which can be clearly resolved as being derived from single protons. As implied earlier, this is very important if we are to be able to differentiate different perturbations in different parts of the molecule, even though the actual part referred to may not be known. The power of this technique can be further enhanced by the use of deconvolution procedures such as those developed by Ernst (1966) (Fig. 1).

The study of these particular resonances has the advantage that their position, which is downfield of the main bulk of the haemoglobin spectrum, is pH-dependent, depending on the state of protonation of the imidazole ring. Thus the pK of the ring protonation provides a further parameter for study as well as providing facility for monitoring other details of the interaction between anaesthetic and protein.

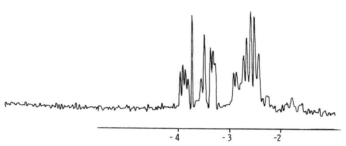

F_IG_. 1. A spin-echo Fourier-Transform nuclear magnetic resonance spectrum of the aromatic-residue region of haemoglobin obtained with mild resolution enhancement by a deconvolution procedure, as used by A. Ferridge (unpublished work, 1977). The left- and right- hand groups of resonances originate from histidyl H-2 and H-4, respectively.

The first n.m.r. experiments with this system (Barker *et al.* 1975) studied the effects of chloroform, halothane, ether and methoxyflurane on haemoglobin by normal Fourier-Transform n.m.r. spectroscopy. We used higher concentrations of anaesthetic than used clinically. In each case, noticeable changes in the chemical shifts of only one or two histidine resonances, out of the 12 titratable ones present, were observed.

More detailed studies by the same methods with halothane (Brown *et al.* 1976) showed that specific perturbations persisted down to clinical concentrations of anaesthetic and we attempted to determine the position in the protein where these effects originated.

The application of the spin-echo techniques has provided a much more powerful method for studying these perturbations, since a slight shift of one of the histidyl residues under observation either towards or away from one of its neighbours might not change the chemical shift much (unless that neighbour was changed) but it could alter its relaxation time considerably. This has proved to be the case. Chloroform, ether, halothane and cyclopropane were studied. These various anaesthetics once again produced a few select perturbations among the histidines, only this time they were manifest not only as changes in chemical shift but also as concentration-dependent fluctuations in intensity. These fluctuations were not, however, proportional to the changes in the chemical shift, as shown in Fig. 2.

The intensity behaviour proved to be a more discriminating perturbation than the chemical shift, although the latter is a prerequisite for this particular pattern of behaviour to be observed. This observed T_2-dependent variation in intensity with anaesthetic concentration is characteristic of the exchange phenomena observed for a fast exchange situation. The equations describing this process have been given by Baldo *et al.* (1975) and the results observed for

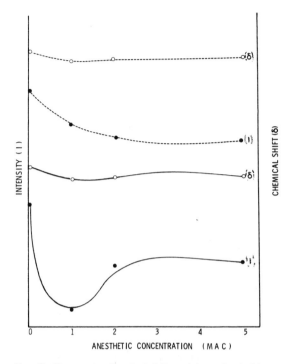

FIG. 2. Changes in chemical shift and intensity (arbitrary units) of two haemoglobin histidyl resonances as a function of diethyl ether concentration (expressed as a multiple of the minimum alveolar concentration, MAC — an index of clinical anaesthetic potency in man). The upper set of lines (---) illustrates the response of most of the histidyl resonances in the presence of anaesthetics. The decay in intensity is probably related to general surface interactions. The lower set of lines (—) illustrates the selective response of only one or two histidyl resonances and the minimum in the intensity/concentration curve is characteristic of an anaesthetic exchange process.

one of the resonances perturbed by several anaesthetics have been fitted to their curve, which describes the broadening of an n.m.r. line as a function of percentage occupancy of a binding site (Fig. 3). The broadening observed is a maximum when slightly over 40% of the binding sites are filled. If the 'binding' properties of each anaesthetic are the same, then the broadening becomes directly related to the molecular concentration of the anaesthetic.

Although the perturbations being described are observed on surface residues, it does not necessarily imply that the interaction is a surface one. Several indications suggest that it is not. It would be surprising if such a non-specific and inert set of compounds as the common anaesthetics were so selective in their interactions with surface histidines that they selected only one or two of all the surface histidines.

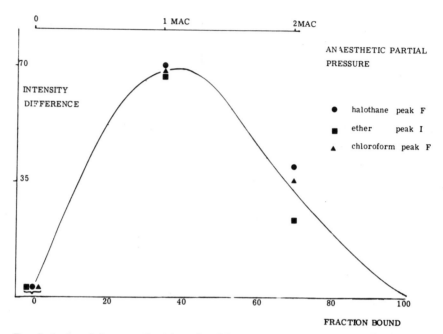

FIG. 3. A plot of the normalized intensity differences between shifted and non-shifted peaks against the percentage fraction of anaesthetic bound (bottom scale) and the anaesthetic partial pressure (upper scale) expressed as multiples of the minimum alveolar concentration (MAC). The solid line is the predicted curve derived from the equation

$$\pi\Delta\nu_{obs} = \frac{1}{T_{2(obs)}} = \frac{f_{H}}{T_2[H]} + \frac{f_{HA}}{T_2[HA]} + f^2_{H}f^2_{HA}(\delta_H - \delta_{HA})^2(\tau_H - \tau_{HA})$$

which describes the fast exchange equilibrium between free haemoglobin (H), the anaesthetic (A) and the complex HA:
H + A ⇌ HA
where $\Delta\nu$ is the line width at half height, τ is the lifetime of a component, δ is the chemical shift of that component, and f is the fraction of that component.

A more plausible conclusion is that the surface perturbations monitor the expansion of hydrophobic pockets within the protein which have been invaded by the hydrophobic anaesthetic molecules. Thus the 'binding' referred to earlier becomes no more than a simple solution process within these pockets and the exchange monitored by the broadening reveals no more than the movement of 'solute' in and out of these pockets. In this case the concentrations which relate to the decrease in T_2 for different anaesthetics would become directly comparable. Fig. 3 demonstrates this as these broadenings turn out to be directly proportional to their anaesthetic potencies and hence to their hydrophobic molecular solubilities.

CONCLUSION

These data provide the first direct evidence that anaesthetics can interact with hydrophobic pockets in proteins and that the site(s) appear to behave as simple bulk solvents in terms of their solubility characteristics. The conformational perturbations produced by such hydrophobic interactions can be transmitted and detected in non-hydrophobic areas of the protein. The correlation of these phenomena with lipid solubility and hence with *in vivo* anaesthetic potency is an encouraging asset for any mechanism of anaesthesia involving protein perturbations (Brown *et al.* 1977). The fact that conformational effects specific to individual agents are also observed in proteins (Barker *et al.* 1975) is relevant to the peripheral mechanisms of the side-effects, biotransformation and toxicity of anaesthetics. Proteins that could be involved in anaesthetic action include enzymes, neurotransmitter receptors, and membrane transport and motility systems. The link between the molecular and cellular effects of anaesthetics may also be mediated through calcium permeability (Allison 1974).

The purpose of the present studies has been to understand more about the 'ground rules' for molecular interactions of anaesthetics. However, the data that we have obtained also indicate that anaesthetics may have a limited role as selective hydrophobic probes for studying protein activity in general.

References

ALLISON, A. C. (1974) The effects of inhalational anaesthetics on proteins, in *Molecular Mechanisms in General Anaesthesia* (Halsey, M. J., Millar, R. A. & Sutton, J. A., eds.), pp. 164-181, Churchill Livingstone, Edinburgh

BALDO, J. H., HALFORD, S. E., PATT, S. L. & SYKES, B. D. (1975) The stepwise binding of small molecules to proteins. Nuclear magnetic resonance and temperature jump studies of the binding of 4-(*N*-acetylaminoglucosyl)-*N*-acetylglucosamine to lysozyme. *Biochemistry 14*, 1893-1899

BALASUBRAMANIAN, D. & WETLAUFER, D. B. (1966) Reversible alteration of the structure of globular proteins by anaesthetic agents. *Proc. Natl. Acad. Sci. U.S.A. 55*, 762-765

BARKER, R. W., BROWN, F. F., DRAKE, R., HALSEY, M J. & RICHARDS, R. E. (1975) Nuclear magnetic resonance studies of anaesthetic interactions with haemoglobin. *Br. J. Anaesth. 47*, 25-29

BRAMMALL, A., BEARD, D. J. & HULANDS, G. H. (1974) Inhalational anaesthetics and their action *in vitro* with glutamate dehydrogenase and other enzymes. *Br. J. Anaesth. 46*, 643-652

BROWN, F. F., HALSEY, M. J. & RICHARDS, R. E. (1976) Halothane interactions with haemoglobin. *Proc. R. Soc. Lond. Sci. 193*, 387-411

BROWN, F. F., HALSEY, M. J. & RICHARDS, R. E. (1977) Protein interactions with general anaesthetics (Abstracts of British Biophysical Society Meeting), in *NMR in Biology* (Dwek, R. A., Campbell, I. D., Richards, R. E. & Williams, R. J. P., eds.), pp. 354-355, Academic Press, London

CAMPBELL, I. D., DOBSON, C. M., WILLIAMS, R. J. P. & WRIGHT, P. E. (1975) Pulse methods for the simplification of protein NMR spectra. *FEBS (Fed. Eur. Biochem. Soc.) Lett. 57*, 96-99

ERNST, R. R. (1966) Sensitivity enhancement in magnetic resonance. *Adv. Magn. Resonance 2,* 1-135

HALSEY, M. J. (1977) Chemical structure and pharmacological action, in *Newer Inhalation Anaesthetics,* Abstracts of Scientific Meeting of the Faculty of Anaesthetics of the Royal College of Surgeons of England, pp. 1-4

HALSEY, M. J. & SMITH, E. B. (1970) Effects of anaesthetics on luminous bacteria. *Nature (Lond.) 227,* 1363-1365

KIESOW, L. A. (1974) Hyperbaric inert gases and the haemoglobin–oxygen equilibrium in red blood cells. *Undersea Biomed. Res. 1,* 29-44

KING, S. M., WHITE, D. C. & ADEY, G. D. (1977) The effects of anaesthetics on some enzymes. *Br. J. Anaesth.,* in press

LAASBERG, L. H. & HEDLEY-WHITE, J. (1971) Optical rotatory dispersion of haemoglobin and polypeptides: effect of halothane. *J. Biol. Chem. 246,* 4886-4893

LEUWENKROON-STROSBERG, E. (1973) *Effect of Volatile Anaesthetics of Protein Structure and Function,* Ph. D. Thesis, Free University, Brussels

MIDDLETON, A. J. (1973) *The Inhibition of Bacterial Luminescence by General Anaesthetics,* D. Phil. Thesis, University of Oxford

MILLAR, R. A., BEARD, D. J. & HULANDS, G. H. (1971) Oxyhaemoglobin dissociation curves *in vitro* with and without the anaesthetics halothane and cyclopropane. *Br. J. Anaesth. 43,* 1003-1011

MILLER, K. W., PATON, W. D. M., SMITH, E. B. & SMITH, R. A. (1972) Physicochemical approaches to the mode of action of general anaesthetics. *Anesthesiology 26,* 339-351

MILLER, K. W., PATON, W. D. M. & SMITH, E. B. (1973) The pressure reversal of general anesthesia and the critical volume hypothesis. *Mol. Pharmacol. 9,* 131-143

NICOLI, M. A., MEIGHEN, E. A. & HASTINGS, J. W. (1974) Bacterial luciferase: chemistry of the reactive sulphydryl. *J. Biol. Chem. 249,* 2385-2392

ÖSTERGREN, G. (1944) Colchicine mitosis, chromosome contractions, narcosis and protein chain folding. *Hereditas 30,* 429-467

SCHOENBORN, B. P. (1965) Binding of xenon to horse haemoglobin. *Nature (Lond.) 208,* 760-762

SCHOENBORN, B. P. (1968) Binding of anesthetics to protein: an x-ray crystallographic investigation. *Fed. Proc. 27,* 999-894

SETTLE, W. (1975) Functional aspects of the interaction of anaesthetics with proteins, in *The Strategy for Future Diving to Depths Greater than 1000 ft* (Halsey, M. J., Settle, W. & Smith, E. B., eds.), pp. 14-18, Undersea Medical Society, Maryland

WEISKOPF, R. B., NISHIMURA, M. & SEVERINGHAUS, J. W. (1971) The absence of an effect of halothane on blood haemoglobin oxygen equilibrium *in vitro. Anesthesiology 35,* 579-581

WHITE, D. C. & DUNDAS, C. R. (1970) Effects of anaesthetics on emission of light by luminous bacteria. *Nature (Lond.) 226,* 456-458

WHITE, D. C., WARDLEY-SMITH, B. & ADEY, G. D. (1975) Anesthetics and bioluminescence. *Progr. Anesthesiol. 1,* 583-591

Discussion

Roberts: Are your data good enough to extract the information on the rate of exchange of the anaesthetic between the bound and free states which can, in principle, be obtained from these changes in line-width?

Halsey: The observed T_2-dependent variation in intensity with anaesthetic concentration is characteristic of a fast-exchange process. We have not yet quantified the rates but have observed, for example, that methoxyflurane

(MeO·CF$_2$·CHCl$_2$) — in contrast to the other clinical anaesthetics so far studied — appears to be in slow exchange with the haemoglobin site(s).

Richards: This technique can yield some interesting structural information about proteins but I do not understand what this has to do with anaesthesia.

Blundell: The binding of halothane by myokinase (Sachsenheimer *et al.* 1977) leads to malignant hyperthermia, characterized by rigidity of muscles and a continuous rise in body temperature.

Halsey: I discussed the general relevance of protein structure and function to anaesthesia but let me make two additional points. First, investigations of anaesthetic effects on proteins form only one aspect of the many current investigations into the mechanisms of general anaesthesia. Other complementary aspects include studies on lipids, cell membranes, neuronal systems and whole animals (Halsey *et al.* 1974). Second, our choice of haemoglobin as the initial model protein for attempting to determine the 'ground rules' for anaesthetic–macromolecule interactions was governed by pragmatic reasons (as I indicated) rather than by its physiological significance in anaesthesia. We are also extending our structural studies both to an example of a membrane-associated protein whose function is already known to be affected by clinical concentrations of anaesthetics (luciferase) and also to a neuroprotein of potential physiological importance (acetylcholine receptor).

Franks: Both you and Professor Richards referred briefly to the effect of pressure on protein stability. Application of pressure to metmyoglobin (at low temperature) unfolds the protein but if, at high pressures, the temperature is increased, then the protein begins to fold again spontaneously (Zipp & Kauzmann 1973). This behaviour reflects the complexity of the hydrophobic effects and points to some connection between the anaesthetic effect and a conformational change which involves the hydrophobic region of the protein.

Halsey: Professor Gutfreund mentioned the pressure-relaxation method for investigating transient conformation changes. What pressures were applied?

Gutfreund: About 200 atm; the effect probably centres on disturbances of charge–charge interactions. We are at present analysing the volume changes from the amplitudes of the various processes. We developed the pressure-relaxation technique to study assembly processes such as in myosin and tubulin, their interaction with other proteins, and regulation of such processes by ligand binding. These processes depend strongly on pressure. Myosin, for example, at high ionic strength dissociates into monomers at a pressure of only 5 atm (Davis & Gutfreund 1976).

Halsey: Are not the volume changes in the whole membrane associated with the lipids greater than those associated with the protein?

Franks: Maybe so, but the existence of a pressure reversal in the unfolding indicates that initially $\Delta V > 0$ but at higher pressures (or temperatures) $\Delta V < 0$. Lipids are compressible but they will not show this reversal of pressure effects.

Halsey: We are referring to 'pressure reversal' in a different sense. The phenomenon observed in animals below 100 atm is that an animal already anaesthetized at 1 atm will wake up as pressure is applied either hydrostatically or with helium (a relatively 'inert' gas in pharmacological terms). This has been termed 'pressure reversal of anaesthesia' (Lever *et al.* 1971). This does not imply that there will be a reversal in sign of the pressure effects on their own. However, it is possible that at pressures above 100 atm the protein-volume-reversal effect might be relevant to the animal studies. For example, current studies on the pressure reversal of intravenous anaesthesia (M. J. Halsey, B. Wardley-Smith & C. J. Green, unpublished observations) indicate that for some agents there is an upper limit to the pressure reversal effect. It has not yet been possible to expose animals to high enough pressures at which the protein reversal effects that you describe could occur.

Gutfreund: Increasing the pressure should lower the permeability of the lipid — rotation of proteins would be inhibited if that depends on the mobility of the lipids. One easy way to find out about the pressure dependence of phase transitions is to measure it calorimetrically.

Franks: The effects of pressure on lipid monolayers are straightforward. A lipid bilayer is probably much the same as two monolayers. But no single instance of a reversal in sign has been observed with lipids, i.e. $(\delta V / \delta P) < 0$.

Halsey: The effects of pressure on their own are interesting physiologically and in particular are related to the high-pressure neurological syndrome, which is one of the major problems in deep-sea diving (Halsey *et al.* 1975). The fact that anaesthetics can ameliorate the syndrome is consistent with the idea that anaesthetics expand the critical molecular sites and pressure contracts them.

Cornforth: Has a chiral form of an anaesthetic of the halogenocarbon type been made? It ought to be possible. Perhaps one could get enantiomers equivalent in all physical properties except chirality.

Halsey: I am not aware of any evidence that chirality is important in general anaesthesia produced by inhalational agents. For example, there is no stereospecific activity of halothane ($CF_3 \cdot CHBrCl$) in an isolated nerve or a phospholipid-bilayer membrane model system (Kendig *et al.* 1973). There are problems in interpreting the stereochemical influences on variations in the pharmacological activities of intravenous anaesthetics but recently significant differences for the ketamine enantiomorphs have been observed in rats (Marietta *et al.* 1977).

Williams: You studied haemoglobin; was it in the carbonmonoxy-haemo-globin form?

Halsey: No, most of our work has been done with unmodified oxyhaemo-globin isolated from human red cells (as described in Brown *et al.* 1976).

Williams: Haemoglobin binds xenon. Is it known where it binds?

Halsey: Schoenborn (1965) studied the binding of xenon to crystals of horse methaemoglobin and identified two possible sites.

Williams: But that is probably where it binds in the deoxy form. As the deoxy form is paramagnetic, I doubt that it would have given n.m.r. spectra like the one you illustrated. So you have eliminated the specific binding problem by using a diamagnetic state and so blocking the normal site for an anaesthetic to bind in haemoglobin.

Halsey: I agree that there may be differences between the anaesthetic effects on the oxy- and deoxy-form (I. D. Campbell and F. F. Brown are studying this at Oxford University, Biochemistry Department). However, any of the anaesthetic sites in haemoglobin may be regarded as 'specific' and capable of being 'blocked' in the oxy-form. If this were so, anaesthetics would significantly shift the shape or position of the oxygen-dissociation curve which has been shown to be insensitive to high concentrations of anaesthetics (Millar *et al.* 1971). There is also no evidence for saturation of the haemoglobin 'solution' sites as the partial pressures of most of the anaesthetics are in-creased. The list of agents studied includes xenon (Schoenborn 1965) and the only known exception is cyclopropane (Gregory & Eger 1968).

Williams: The methyl resonances in the n.m.r. spectrum might be good indicators of the specific effects in the deoxy form.

Halsey: The methyl resonances associated with the haem group of ferri-myoglobin cyanide have been studied by Schulman *et al.* (1970) who observed various shifts produced by xenon and cyclopropane. Unfortunately, they used a single relatively-high concentration of the two agents. Comparison between these experiments and other anaesthetic effects on proteins is difficult.

Lipscomb: The binding-site of xenon in haemoglobin is far removed from the haem but it is near the haem in myoglobin (see Schoenborn 1965).

Knowles: But, to echo Professor Richards' point, what is the relevance of this? Correlations can be drawn between all sorts of parameters: the birth rate in England during this century correlates splendidly with the population of storks in Heligoland!

Halsey: Maybe our attempt to establish ground rules for the interaction between proteins and anaesthetics was overambitious but until recently few people believed that anaesthetics ever affected proteins.

Lipscomb: If there is a general effect on proteins, do you observe an effect in a simple model system like bacterial motility?

Halsey: There are anaesthetic effects on bacterial proteins such as luciferase and also on various motile proteins (for a general review see Allison 1974). Most studies have been functional rather than structural and what is now needed is clear understanding of the link between the effects.

References

ALLISON, A. C. (1974) The effects of anaesthetics on proteins, in *Molecular Mechanisms in General Anaesthesia* (Halsey, M. J., Millar, R. A. & Sutton, J. A., eds.), Churchill Livingstone, Edinburgh

BROWN, F. F., HALSEY, M. J. & RICHARDS, R. E. (1976) Halothane interactions with haemoglobin. *Proc. R. Soc. Lond. B 193*, 387-411

DAVIS, J. S. & GUTFREUND, H. (1976) The scope of moderate pressure changes for kinetic and equilibrium studies of biochemical systems. *FEBS (Fed. Eur. Biochem. Soc.) Lett. 72*, 199-207

GREGORY, G. A. & EGER, E. I. JR. (1968) Partition coefficients and blood fractions at various concentrations of cyclopropane. *Fed. Proc. 27*, 705

HALSEY, M. J., MILLAR, R. A. & SUTTON, J. A. (eds.) (1974) *Molecular Mechanisms in General Anaesthesia*, Churchill Livingstone, Edinburgh

HALSEY, M. J., SETTLE, W. & SMITH, E. B. (eds.) (1975) *The Strategy for Future Diving to Depths Greater than 1000 ft*, Undersea Medical Society, Maryland

KENDIG, J. J., TRUDELL, J. R. & COHEN, E. N. (1973) Halothane stereoisomers: lack of stereospecificity in two model systems. *Anesthesiology 39*, 518-524

LEVER, M. J., MILLER, K. W., PATON, W. D. M. & SMITH, E. B. (1971) Pressure reversal of anaestheia. *Nature (Lond.) 231*, 368-371

MARIETTA, M. P., WAY, W. L., CASTAGNOLI, N. & TREVOR, A. J. (1977) On the pharmacology of the ketamine enantiomorphs in the rat. *J. Pharmacol. Exp. Ther. 202*, 157-165

MILLAR, R. A., BEARD, D. J. & HULANDS, G. H. (1971) Oxyhaemoglobin dissociation curves *in vitro* with and without the anaesthetics halothane and cyclopropane. *Br. J. Anaesth. 43*, 1003-1011

SACHSENHEIMER, W., PAI, E. F., SCHULTZ, G. E. & SCHIRMER, R. H. (1977) Halothane binds in the adenine-specific niche of crystalline adenylate kinase. *FEBS (Fed. Eur. Biochem. Soc.) Lett. 79*, 310-313

SCHOENBORN, B. P. (1965) Binding of xenon to horse haemoglobin. *Nature (Lond.) 208*, 760-762

SCHULMAN, R. G., PEISACH, J. & WYLUDA, B. J. (1970) Effects of cyclopropane and xenon upon the high resolution nuclear magnetic resonance spectrum of ferrimyoglobin cyanide. *J. Mol. Biol. 48*, 517-523

ZIPP, A. & KAUZMANN, W. (1973) Pressure denaturation of metmyoglobin. *Biochemistry 12*, 4217-4221

Physical and chemical properties of lysozyme

C. C. F. BLAKE, D. E. P. GRACE, L. N. JOHNSON, S. J. PERKINS & D. C. PHILLIPS

Laboratory of Molecular Biophysics, Department of Zoology, University of Oxford, Oxford

R. CASSELS, C. M. DOBSON, F. M. POULSEN & R. J. P. WILLIAMS

Inorganic Chemistry Laboratory, University of Oxford, Oxford

Abstract   The conformations of lysozyme in crystals and in aqueous solution are discussed and it is shown that the basic conformation is similar in the two states. Certain parts of the molecule have mobility. The reactions of lysozyme with protons, metal ions and some organic reagents are examined in the light of the conformations and their dynamics. The reactions considered are mainly those of tyrosyl, tryptophyl and carboxylate residues. The reactivity data are used in a discussion of the energy states of the reacting side-chains. In particular the reactivity of Glu-35 and its interaction with Trp-108 lead to suggestions for some new aspects in the hypothesis for the mechanism of action of lysozyme. In most respects the X-ray crystal diffraction and the nuclear magnetic resonance solution studies are in accord and complementary.

We note at the outset that, for a convincing discussion of environmental effects due to protein structure on the reactivity of side-chains, the crystal state structure must correlate well with the solution state structure. This necessity arises from the facts that conformations are best defined in the crystal or by working from knowledge gained from crystal studies and reactivities are usually and more meaningfully analysed in solution. Any differences between crystal and solution conformations, if they exist, could be critical in an understanding of reactivity. Lysozyme provides the first opportunity for a full analysis of what is meant by the environment of a group in solution conditions of reaction as it is the first enzyme for which we have a sufficiently good correlation between solution and solid-state structural data.

To familiarize readers with the lysozyme structure we start with a general survey of the molecule in the crystalline state and shall then refer to the correlation between this conformation and a detailed examination of the nuclear magnetic resonance (n.m.r.) spectrum in solution. Although previously many methods have been used in solution studies, only n.m.r. carries sufficiently detailed and precise conformational information to allow a

close comparison with the X-ray structure (Dobson 1977). With certain
provisos we conclude that the conformation observed in solution is the same
as that which is seen in the crystal. New observations on the mobility of the
structure are reported.

In the light of firm knowledge of the solution conformations and their
mobilities we can re-examine the reactions of the protein. It is here that n.m.r.
is such a powerful technique since once all the signals arising from a specified
type of side-chain have been assigned one can follow the changes in each one
of these side-chains unequivocally even though the protein is in solution. The
effects of the reversible binding of protons, metal ions or other molecules on
individual side-chains can be established unambiguously. In the past this has
not been possible and chemical inferences have had to be drawn in order to
identify the residues concerned in reversible changes. In the case of irreversible changes due to chemical attack, it has been possible to identify the sites of
attack through the analysis of fragments of modified sequence. N.m.r. gives
the same information about the location of an irreversible attack but gives
additional information about the conformational modifications in the protein
resulting from the attack. Degradative analysis cannot yield these data but an
X-ray structure determination yields parallel data in the crystalline state to the
n.m.r. observations in solution. In this article we shall confine ourselves to a
discussion of reactions which we have followed by n.m.r. spectroscopy.

While attempting to explain our observations we want to avoid if possible
reference to bulk physical constants to explain reactivity since we believe that
we have now reached the stage at which we ought to describe reactivities
observed in solution in much more specific chemical terms. Thus although it
can be said that the reaction mechanism of lysozyme has already received a
good general description, our new microscopic physical and chemical inspection of the molecule shows that there may be further important considerations related to the enzyme activity which have been passed over. Our
approach starts from an effort to understand the chemical environmental
factors that can come into play in the enzyme from simple studies of the
binding of protons and metal ions before we look at the *binding* of inhibitors
and substrates. Subsequently we shall look at the chemical environmental
factors which could affect simple chemical *attack* by organic reagents on the
protein side-chains before we consider the possible intermediates and transition states of the substrate *reaction*. We shall show that we must understand
further both the ground states of parts of the molecule and better appreciate
the role of group-motions in chemical reactions before we can reach a final
conclusion about how this enzyme works. We also believe that this is true of
other enzymes.

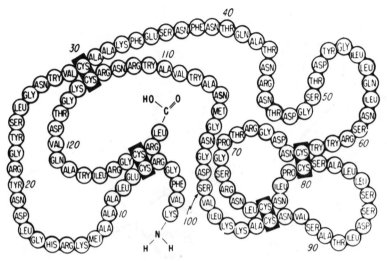

FIG. 1. Amino-acid sequence of hen egg-white lysozyme (after Canfield & Liu 1965). Residue 103 is Asn (see Imoto *et al.* 1972).

STRUCTURE OF LYSOZYME

Crystallographic evidence

The lysozyme molecule: Hen egg-white lysozyme (EC 3.2.1.17) is a relatively small and stable enzyme molecule with a single polypeptide chain of 129 amino-acid residues cross-linked by four disulphide bridges. The amino-acid sequence (Fig. 1) was determined by Canfield (1963) and Jollès *et al.* (1963); Blake *et al.* (1965, 1967a) established the three-dimensional structure, in tetragonal crystals at pH 4.7 with 1M-sodium chloride, and acetate buffer. More recent studies of the enzyme isolated from humans (Banyard *et al.* 1974), turkey (Bott & Sarma 1976) and tortoise eggs (Blake *et al.*, personal communication) have shown that the conformation of the molecule is highly conserved in these different species and crystal structures.

The main features of the three-dimensional structure are shown in Fig. 2. The molecule has roughly the shape of a prolate spheroid with overall dimensions about $45 \times 30 \times 30$ Å^3 and a deep cleft on one side which is the active site of the enzyme. The polypeptide chain is folded on itself so that the first 40 residues from the NH$_2$-terminal form a compact globular region with a hydrophobic core trapped between two α-helices (residues 4–15 and 24–35). Residues 40–85 are more generally hydrophilic in character (Phillips 1967) and fold to form a second domain composed partly of a β-pleated sheet, one

FIG. 2. Perspective drawing of the conformation of the main polypeptide chain in hen egg-white lysozyme. Only positions of α-carbon atoms are shown.

surface of which is part of the outer surface of the complete molecule and the other lines one side of the active-site cleft. The remainder of the polypeptide chain partially fills the gap between the two domains with a helix and then folds around the NH_2-terminal domain, building up the other side of the active-site cleft mainly with relatively hydrophobic residues. Overall some 59 residues are in α-helices and 20 in β-pleated sheets. Three water molecules are trapped within the protein structure (Imoto *et al.* 1972).

The distribution of the amino-acid residues in this three-dimensional structure has been studied in detail, especially with regard to the density with which the various groups of atoms are packed within the molecule (Richards 1974) and the differing degrees to which they are exposed to the surrounding liquid (Lee & Richards 1971). The packing density observed in this molecule generally approaches that found in crystals of organic compounds but there are some variations with position in the molecule. Thus the packing density is high in the so-called hydrophobic box (Blake *et al.* 1965) formed by the

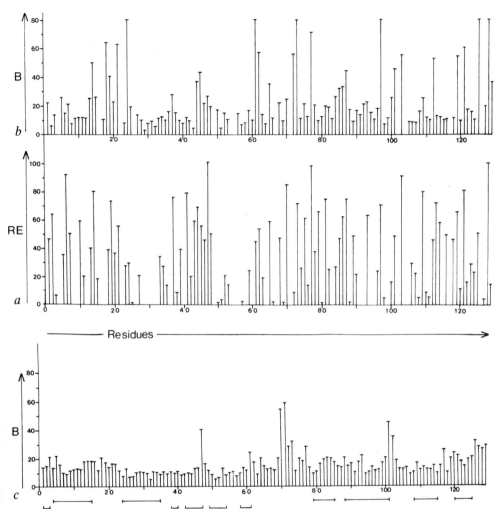

FIG. 3. (a) Relativé exposures of amino-acid residues on hen egg-white lysozyme (Lee & Richards 1971); (b) B_i values (proportional to mean-square displacements from mean positions) for side-chain atoms at the present stage of refinement of the crystal structure; (c) B_i values for main-chain atoms.

side-chains of Tyr-23, Trp-28, Trp-111 and Trp-108 around Met-105 (see later). This hydrophobic box is much altered in human lysozyme (Banyard *et al.* 1974). In the region between the side-chains of Leu-8, Met-12, Ala-32, Ile-55, Leu-56, Ile-88 and Val-92 there is a cavity in the structure and the packing density is correspondingly low. The results of the study by Lee & Richards (1971) of the relative exposures of the amino-acid side-chains are summarized in Fig. 3a, which shows the relative exposure as the surface area of the side-chain atoms accessible to water in the lysozyme molecule divided by the surface area of the same side-chain that is accessible to water in a standard reference conformation. The side-chains that are completely buried (e.g. Met-105, Trp-28) are immediately recognizable as having zero relative exposure. Similar results were obtained by Shrake & Rupley (1973).

Motion in the crystalline state: thermal motion. Work now in progress is aimed at refining the description of the crystal structure by minimizing the differences between the X-ray observations and the corresponding quantities calculated from the parameters describing the structure. The conventional agreement index, $R = \Sigma |F_o - F_c| / \Sigma F_o$ (where F_o and F_c are the observed and calculated structure amplitudes, respectively), is 0.22 at present and further improvement is expected. This refinement of the structure has involved generally small adjustments to the positions of the atoms in the model of the protein molecule, the location of most of the water molecules in the crystal, and the inclusion of isotropic temperature factors to take into account the thermal motion and disorder of the protein and water molecules. Establishment of the position of solvent molecules, many of which appear to be ordered on the time scale of the X-ray measurements, and assessment of the significance of such temperature factors are among the most difficult problems that arise in the refinement of protein structures but the results now obtained seem promising and merit preliminary discussion especially in comparison with the n.m.r. observations. Probable water positions are referred to below and we consider here the X-ray evidence for molecular motion.

The contribution of each atom to a particular structure amplitude is multiplied by a temperature factor $\exp(-B_i \sin^2\theta/\lambda^2)$, where θ is the Bragg angle through which the X-rays of wavelength λ are reflected and B_i is a constant which, for pure harmonic thermal vibrations, is related to $\bar{u}_i^2$, the mean square amplitude of vibration of the atom in question, by the formula $B_i = 8\pi^2\bar{u}_i^2$. Thus when $B_i = 80$ Å^2 the corresponding atom may be taken to be vibrating with a mean square amplitude of about 1 Å^2.

In order to keep the number of independent parameters used to describe the crystal structure as small as possible in comparison with the number of X-ray observations (about 9600 structure amplitudes at 2 Å resolution) we used only two different B_i values to describe the 'thermal motion' of each amino-acid residue. The first of these has been applied to all of the atoms contributing to the main polypeptide chain (N, Cα, C' and O) and to the Cβ atom when there is one. The second B_i value is used for all the side-chain atoms beyond Cβ. The values obtained at the present stage of the analysis are shown diagrammatically in Fig. 3b and c.

Among the values of B for main-chain atoms three groups stand out. Ser-47 has the most exposed side-chain and the main-chain density for this residue was weak in the original electron-density map determined by isomorphous replacement. The main-chain atoms here may be vibrating with large amplitudes (even though neighbouring groups are relatively still) but it is more probable that two or more different conformations are represented, with the atoms in each vibrating to a lesser extent. Residues 70–73 were difficult to locate in the original map in an unstrained conformation (Levitt 1974) and the refinement of the structure of lysozyme in triclinic crystals (L. H. Jensen, personal communication) has shown this part of the molecule to be in a somewhat different conformation from the one considered here. It seems probable that a complete description of the structure will have to take into account alternative conformations in this part of the molecule which is, incidentally, a principal antigenic determinant (Shinka *et al.* 1967; Arnon & Sela 1969). Finally, the dipeptide Asp-Gly (101-102) is located on an exposed loop and it is impossible at this stage to decide whether the high B values represent thermal motion of high amplitude or several alternative and differently populated conformations.

Apart from these observations which point to the probability that parts of the molecule vibrate differentially, the general variation of the B values for main-chain atoms seems consistent with the hypothesis that to a first approximation the lysozyme molecule in the crystal moves as a rigid body partly in vibrational and partly in librational motion. Thus it is clear in Fig. 3*c* that the B values are least for the buried residues in the centre of the molecule and tend to increase towards the carboxy end of the chain which, as noted before, is on the surface of the molecule. In a first rough analysis of the data we plotted (Fig. 4) the values of B_i for the main-chain atoms of individual residues against the squares of their distances from the molecular centroid. Despite the large variations, the plot suggests that this simple model, with the purely vibrational motion equivalent to a B value of about 7 Å^2, is a promising one that merits further elaboration. Further refinement and analysis may make possible a more detailed description of the molecular motion including the identification of intramolecular modes of vibration of the kind considered by McCammon *et al.* (1976).

The B values for the side-chains shown in Fig. 3*b* generally correlate well with the relative exposure figures shown in Fig. 3*a*. Small B values are associated with buried side-chains and high B values with exposed side-chains, though it must be noted that not all exposed side-chains have high B values. Detailed analysis will need to take into account the intermolecular contacts in the crystals as well as the properties of the individual molecules.

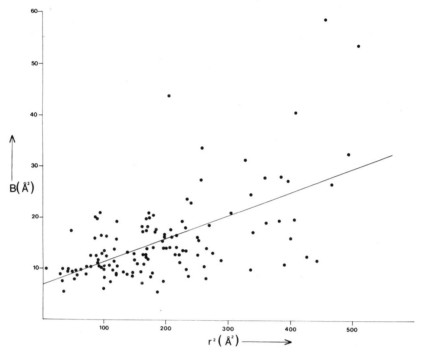

FIG. 4. B_i values for the main-chain atoms of individual amino-acid residues plotted against the squares of their distances from the molecular centroid. The best regression line is $B = 6.9 + 0.045r^2$ with a highly significant correlation coefficient of 0.6.

Two side-chains are particularly worthy of note. Trp-62 was represented in the original map by good density only for the five-membered ring of the indole and this was taken to indicate librational motion about the Cβ–Cγ bond. Similarly the density for the Cγ atoms of Val-109 was weak and embraced the three staggered positions available to the Cγ atoms by rotation about the α–β bond. Other side-chains (e.g. Arg-73, Arg-125, Arg-128) were poorly represented in the original electron-density map and may adopt different conformations in different molecules besides being in extreme thermal motion. No account has been taken at this stage of variations in B values along side-chains although there is good evidence for such variations in many instances in addition to Trp-62.

The above observations are now compared with the n.m.r. evidence for molecular conformation and motion.

N.m.r. description of structure and mobility

We shall not describe here the general procedures which we have used both to resolve and assign the ^{1}H n.m.r. spectrum of lysozyme (Dobson 1977). It is sufficient to note that the analysis of ring-current shifts, of shifts due to Pr(III) binding, of changes in relaxation rates of resonances due to Gd(III) binding, and of local, inter-side-chain, proton–proton interaction and nuclear Over-hauser studies gives over 100 vector or scalar quantities which can be used for comparison of the X-ray structure given above with data collected for the solution structure. The n.m.r. mapping procedures give information about local regions of the protein spreading some 5–20 Å from a given probe centre. The procedure is then totally distinct from the diffraction method. The overall quantitative agreement between the n.m.r. derived vectors and scalars and the same constants read off from the model prepared from the X-ray data allow us to conclude with certainty that the overall spatial disposition of amino-acid side-chains in the solid and solution states is very similar. The main-chain fold of the protein in the crystal structure must be maintained in solution. The accuracy of the crystallographic data can be clearly stated but at present the n.m.r. observations are less easily interpreted, and variation in side-chain atomic positions of ± 2 Å cannot be excluded. For many side-chains higher accuracy than this cannot be expected since thermal motions in lysozyme in solution are marked. The surface of lysozyme has high mobility which is observed in the rapid relaxation of lysine side-chain resonances. There cannot be a firm hydration sphere in solution. All three tyrosines flip rapidly; all methyl groups rotate, some valines, leucines and isoleucines flip, probably all phenylalanines flip, and at least one tryptophan (62 or 63) flaps. These motions mean that n.m.r. data must be interpreted as being derived from a sum of locally-different conformations and this has been done, where possible.

The temperature dependence of the structure in solution shows (Fig. 5) that the protein as a whole expands with a small coefficient of expansion, as seen by the gradual changes of ring-current-shifted resonances back to their un-shifted positions. The n.m.r. measurements have not yet revealed local high mobility of any region of the main chain. Some such motions must occur if the tyrosine rings are to flip.

The overall conclusion that the conformation of lysozyme is little affected by change of phase remains true but the observed variations in mobility of groups within the structure especially locally must be kept in mind in any discussion of reactivity. The general study of peptide N–H exchanges in lysozyme (i.e. ^{2}H/^{1}H exchange) previously gave less specific knowledge of the looseness of some parts of the structure although others were apparently much firmer (for references see Imoto *et al.* 1972). Lysozyme does not conform to the type of protein visualized by Linderstrøm-Lang & Schellman (1959) since the mobility is largely restricted to side-chains and not to the main chain.

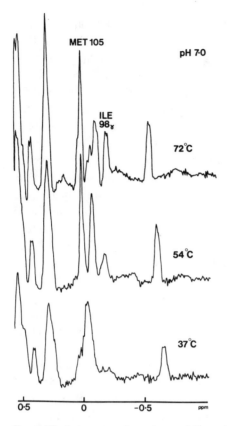

FIG. 5. The temperature dependence of the methyl region of the ^{1}H n.m.r. spectrum of lysozyme. The outermost ring-current shifted resonances collapse slowly at higher temperatures. Other resonances sharpen. Both changes indicate changes in motional freedom in the molecule.

New n.m.r. methods. Aue *et al.* (1976) and Freeman & Hill (1975) have described two-dimensional n.m.r. spectroscopy which will greatly assist resolution and assignment. Their method has been applied to lysozyme (Campbell & Dobson 1978) and to the trypsin inhibitor protein (Nagayama *et al.* 1977; see also this volume p. 179). Further improvement in mapping is available through unpublished work by E. M. Bradbury & G. Chapman at Portsmouth and Dobson, Campbell, Ratcliffe & Williams (1978) at Oxford using methods in which difference spectra between a normal and a selectively-pulsed spectrum are collected continuously. In the accumulated difference spectrum the method gives only those resonances which are close enough in space to the irradiated nucleus to be relaxed by it. The method could be looked upon as a use of local spin interactions and has as its basis the nuclear Overhauser effect.

TABLE 1

Range of conditions included in the studies

Variable	Crystallographic			N.m.r.
	Hen egg-white lysozyme		Human lysozyme	
	Tetragonal	Triclinic	Orthorhombic	
Water content	34%	26%	33%	The usual concentration used is in the range 1–5 mmol/l (i.e. in the range 15–75 g(l)
pH	2.6,4.7,7.2	4.5	4.5	3–11
Salt	1M-NaCl	0.24M-NaNO$_3$	7M-NH$_4$NO$_3$	Low ionic strength to 1.0M-NaCl, varying concentrations of LnCl$_3$, CaCl$_2$ etc.
Temperature	Room temperature, 20 °C			0–75 °C
Specific binding agents	Metal ions, inhibitors, anions			Metal ions, inhibitors, anions

Together with the use of earlier procedures it can be stated that the n.m.r. methods for protein studies now give detailed information on both the conformation and the dynamics of the conformation, and one can in principle obtain a protein structure without resort to a model determined in the crystalline state.

It is against this background that we shall now summarize the observations which we have made by X-ray diffraction and n.m.r. spectroscopy on the environmental influences on reactivity in this protein. We expect these to appear strongly where there is a special surface, that is in the groove of the protein (Fig. 2) rather than elsewhere. This follows not just from general experience of protein (enzyme) chemistry, but from the obviously special energetics of groups in a groove as opposed to any other region of a structure. However, a protein is a cooperative unit and as such it may well be that the special energy states of individual groups are linked to the energy of the whole protein fold. The cooperativity is not just concerned with the energetics of particular conformations but also covers the range and the energies of conformational variation — the dynamics of the structure. We shall seek knowledge of both energy and dynamics in local protein regions. As both are likely to depend on the conditions of measurement we have studied the enzyme in various conditions (Table 1).

TABLE 2

Species variations of aromatic and carboxylic acid residues (number of sequence differences from hen egg-white lysozyme is given in parentheses)

Hen egg-white lysozyme[a]	Chachalaca lysozyme (27)[b]	Human lysozyme (52)[a]	Rat lysozyme (58)[c]
Phe-3	Tyr		Tyr
Phe-34	Tyr	Trp	His
Phe-38	Tyr	Tyr	Tyr
His-15	Tyr	Leu	Asn
Tyr-20			
Tyr-23		Ile	Val
Tyr-53			His
Trp-28			
Trp-62		Tyr	Tyr
Trp-63			
Trp-108			
Trp-111			
Trp-123		Tyr	Tyr
Asp-18			Ser
Asp-48	Asn		
Asp-52			
Asp-66			
Asp-87		Asn	
Asp-101			
Asp-119			
Glu-7			
Glu-35			

[a] See Imoto *et al.* (1972)
[b] See Jollès *et al.* (1976)
[c] See White *et al.* (1977)

METHOD OF INSPECTING ENVIRONMENT AND REACTIVITY

Once the environment of the residues and their changes with conditions are defined in space we need to know clearly the consequences of the structure in terms of energetics. The energy states of the system are open to examination by spectroscopy and again n.m.r. is a standard method for the study of ground state energies; u.v., visible and i.r. absorption spectra help to define electronic and vibrational excited states. The reactivities of groups have to be tackled by chemistry and we hope they will be understood in terms of their ground- and

TABLE 3

Positions of residues seen from the front (as in Fig. 2)

Left-hand side of groove	Trp-62, Trp-63, Asn-59, Asp-52, Asn-44
Right-hand side of groove	Asp-101, Ala-107, Val-109, Ala-110, (Glu-35), Arg-114, (Trp-108)
Back of groove	Ile-98, Ile-58, Leu-56, Gln-57, (Trp-108), Glu-35, (Phe-34), Ser-36
Hydrophobic core	Trp-28, Trp-111
Surface (in part)	Tyr-20, Tyr-23 (Trp-63), (Trp-123), (Trp-108)
Surface	Asp-101, Trp-62

excited-state environments and energies. We shall use two major types of reaction:

(1) *equilibrium* studies of proton, metal, inhibitor or substrate binding, many of the effects of which can be followed by X-ray diffraction as well as spectroscopic techniques;

(2) *kinetics* of organic modification or of the on/off binding of protons (N-H) or of substrates and inhibitors. The products can be examined by X-ray diffraction. In the past definite knowledge could not be obtained about such reactions since chemical inference always had to be used (see Imoto *et al.* 1972). A special case is provided by a biological substitution and it is important to note which amino acids are conserved (Table 2) and the effects of changes of the amino acids on reactivity going from species to species.

COMPARISON OF SIDE-CHAIN REACTIVITIES AND ENVIRONMENTS

The groups that we shall discuss in greatest detail lie in the groove of lysozyme but as these groups show some unexpected features we must also look at groups which are in less important regions, for instance, on the back surface of the molecule or in the core. Table 3 lists the locations of the residues with which we shall be concerned for the most part.

Let us describe first an example of behaviour which does not require special discussion. Histidine-15 is at the back of the molecule (see Fig. 2) on the surface, where it acts as donor in a hydrogen bond with Thr-89, which in turn acts as a donor to Asp-87. Arg-14 and Lys-96 are also in the vicinity. Perhaps for this reason His-15 has a somewhat low pK_a (5.2–5.8) but in general its properties are closely characteristic of histidine in solution. It is not a conserved residue and chemical modification does not affect conformation elsewhere in the protein or catalytic action. A second but different example is tyrosine-53 which is locked into the protein (left-hand side) by the pleated-sheet fold, behaves as an inaccessible group and shows no pK_a in the expected pH range and little chemical reactivity. This is again a straight-forward environmental effect as far as reactivity is concerned.

Many residues in the groove are hydrophobic and in the extreme cases have simple alkyl side-chains. Examples are:

(1*a*) Hydrophobic rings: Trp-62, Trp-108, (Phe-34)

(1*b*) Aliphatic chains: Leu-75, Val-99, Ile-98, Ala-107, Val-109, Leu-56, Ile-58

Residues near or in the groove also include some polar groups:

(2) Polar residues: Glu-35, Asp-52, Asp-101, Arg-114, Asn-44, Asn-37, Ser-36, Asn-59

Although it is possible to discuss the functional roles of residues in classes (1*a*) and (2) on the basis of chemical reactivity, it is often difficult to discern the functional role of residues in class (1*b*) since they have such low reactivity. We cannot dismiss them as of no importance, however, for they can provide steric specificity to a reaction and, largely through repulsive interactions, could control the reaction energetics. For example, there are now examples in proteins where the physical onset of flipping of an aromatic residue costs more than 90 kJ/mol which is numerically large enough to dominate the activation energy of an enzyme reaction (Williams 1977). Simple dynamics of inert side-chains can no longer be ignored in mechanistic thinking. Both X-ray diffraction and n.m.r. data show that Val-109 has considerable rotational freedom.

We shall choose for internal comparison of reactivity two different classes of side-chain: carboxylate groups and aromatic groups. We shall compare the carboxylate groups aspartate-101 and -52 and glutamate-35 and the aromatic groups tyrosine-20, -23 and -53 and tryptophan-28, -62, -63, -108, -111 and -123. We shall describe the environments of these different residues in detail before we attempt to describe their activities on the basis of these environments.

Tyrosine residues

Crystallographic evidence for tyrosine environments

(*1*) *Tyrosine-20:* Tyr-20 (relative exposure, RE, 0.37) is the most exposed of the tyrosine residues. Its phenolic ring lies in a shallow depression on the protein surface with the hydroxy group in contact with the hydroxy group of Ser-100 (3.58 Å away) and with some five water molecules within 4 Å (the closest at 2.90 Å). One face of the ring lies against the aliphatic chain of Lys-96 with one edge in contact with the side-chain of Val-99 which it shields from the surrounding liquid. The other edge is slightly tilted away from the protein and is exposed to solvent with the amino group of Lys-96 about 5.23 Å from $C\varepsilon_1$. The guanidinium group of Arg-21 ($N\varepsilon$) is about 5.26 Å from the edge ($C\varepsilon_2$). In the crystal structure there is no group from neighbouring lysozyme molecules within 4 Å of this residue and we presume that the structure around it will be maintained in solution.

(2) *Tyrosine-23:* This residue also lies in a depression on the protein surface (RE 0.28) and forms one side of the 'hydrophobic box' enclosing Met-105. The Tyr-23 side-chain lies between the basic side-chains of Lys-116 and Arg-31 and the intramolecular contact distances (7.97 and 7.62 Å, respectively, from $C\varepsilon_2$ and $C\varepsilon_1$) that are observed in the crystal structure are longer than they may be in free solution. One face of the ring is involved in an intermolecular contact in the crystal (Blake *et al.* 1967a, Fig. 13): the guanidinium group of Arg-114 in a neighbouring molecule is about 3.4 Å from $C\varepsilon_1$.

(3) *Tyrosine-53:* This residue is more deeply buried in the protein surface (RE 0.14) where it is closely coordinated with other residues and lies with its phenolic ring roughly parallel to the antiparallel β-pleated sheet formed by residues 41–54 and opposite to the side-chain of Asp-52 (cf. Fig. 7). Thus, for example, $C\delta_2$ is close to the carbonyl carbon atom of Asp-52 (3.46 Å), the carbonyl oxygen atom of Thr-51 (3.22 Å) and Cβ of Cys-80 (3.61 Å). The hydroxy group is in close contact with a carboxyl oxygen atom of Asp-66 (2.59 Å) and with a water molecule (2.72 Å) and it is somewhat distantly in contact with the hydroxy group of Thr-51 (3.93 Å). This residue is not directly involved in intermolecular contacts in the crystals. For the flipping observed in solution to occur it would appear to be necessary for the surrounding protein groups to vibrate cooperatively.

N.mr. evidence. In solution we must follow the reactions of these three tyrosines individually. The n.m.r. studies enable us to do this as each tyrosine has been assigned and its *ortho-* and *meta*-protons have been recognized in the ^1H n.m.r. spectrum (Fig. 6). Each tyrosine resonance gives rise to two two-proton doublets. This fine structure of the tyrosine resonances shows that all three tyrosines flip at rates greater than 10^2 s^{-1}. The positions of the resonances are not as expected for free tyrosine but are approximately consistent with ring-current shifts from the neighbouring aromatic rings of residues close to the tyrosines which are observed in the crystallographic study. We shall

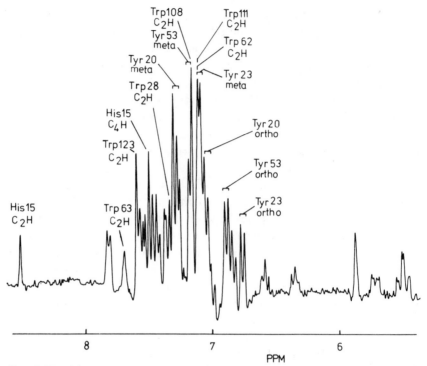

FIG. 6. The C-¹H n.m.r. spectrum of lysozyme showing the two two-proton doublets of each assigned tyrosine, the histidine singlets, and the C(2)-H protons of the six individual tryptophans.

assume that the crystal structure gives a good basis for the discussion of reactivity with the added knowledge that the surface residues (e.g., lysines around tyrosines-20 and -23) are freely mobile and that the tyrosines themselves have considerable mobility. The observed mobility of tyrosine-53 requires some breathing of the protein groups close to it.

Tyrosine reactivities. Simple pH titration following all the resonances by n.m.r. shows definitively that Tyr-23 has a pK_a of about 10.5 but Tyr-20 and -53 do not ionize below pH 11.0. It is difficult to explain this difference between Tyr-23 and Tyr-20. Again, it is Tyr-23 which reacts readily with many reagents for phenols; it is attacked by I_3^- and chloronitrofuroxans (NBF). There is nothing unusual about the occurrence of such reactions in that they represent just a change of 'solvent' and accessibility of the phenol from Tyr-53 (more non-aqueous) to Tyr-23 (partly aqueous). The reasons for the attack on Tyr-23 rather than Tyr-20 are not so clear. Tyr-53, Tyr-20 and Tyr-23 have

TABLE 4

Effect of modifications to tyrosine-25 on n.m.r. signals[a]

Modification	Resonance affected
Iodination	Trp-28, C(2)-H
NBF-treatment	Met-105, S-CH$_3$
	Trp-28, C(2)-H
	Trp-28, N(1)-H
	Trp-111, C(2)-H

[a] The n.m.r. data given in this table and others later in the article are only qualitative indications of changes in the proteins.

comparable flipping motions. Tyr-23, as well as having a lower pK_a, has an environment of positive charge (see above) but in this respect it is not obviously different from Tyr-20. We have examined the structure in the hope of finding a site for I$_3^-$ near Tyr-23 which has characteristics such that I$_3^-$ attack was facilitated (i.e., a well adapted positively charged pocket which would form an intermediate complex). We could not find such a site. A full explanation needs a description of the stabilization of the transition states and chemically transformed intermediates along the reaction path as well as of any such initial complex.

The effect of the two chemical modifications of Tyr-23 on the protein as a whole is described in more detail in Table 4. The effects of formation of an iodinated phenol can be interpreted as ring-current changes within the Tyr-23, or small movements of this ring, influencing the shifts on Met-105 and Trp-28. All the residues are in the hydrophobic box. The effect of the formation of an NBF derivative is somewhat bigger and extends further into the hydrophobic box. It could be that the conformation of the box is somewhat altered by these changes (see Table 4 and Fig. 2). (The hydrophobic box seems to be a likely path for transfer of information within the protein; see later.) There are as yet no X-ray diffraction data on the changes caused by these reactions. We can conclude the examination of the tyrosines by stating that there is no major anomaly in their reactivities which are broadly consistent with their environments though they have not yet been understood in detail.

The reaction of tyrosine-23 with NBF is of further interest, however, since Ferguson et al. (1974) observed that a similar reaction occurred with only one of the three subunits of F$_1$ in ATP synthetase. Moreover, this reaction was followed by a transfer of the reagent from a tyrosine to a lysine on increasing the pH. A similar reaction occurs in lysozyme. In ATP synthetase the question that arises is why one subunit has a reactive tyrosine while the others do not?

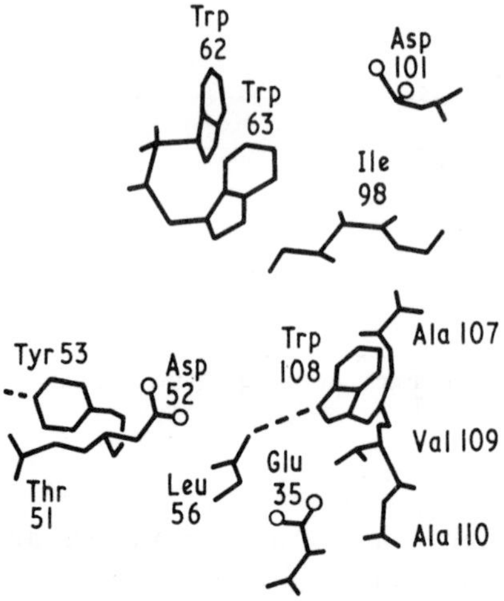

FIG. 7. A semi-schematic representation of the upper part of the groove of lysozyme (see Fig. 2).

In lysozyme the comparable question is, which lysine receives the NBF from the tyrosine, as none is close in the X-ray crystal structure? The most likely candidate is lysine-116. It remains to be established whether the reaction in lysozyme is intra- or inter-molecular. (We thank Dr S. Ferguson, Oxford, for discussing these points with us.)

Tryptophans

Hen egg-white lysozyme is remarkable in having six tryptophan residues, three of which (62, 63, 108) are in the active-site cleft (Fig. 7). The relative exposures of these residues are shown in Table 5 as well as Fig. 3. Four of these tryptophan groups are conserved in human lysozyme (Table 2) and it is the two most exposed (62 and 123) that are changed (to tyrosine) even though 62 is located in the active-site region.

Crystallographic evidence of tryptophan environments

(1) Tryptophans-28, -111 and -123: Of the tryptophan groups Trp-28 is completely buried in the hydrophobic core of the protein where it forms the inner wall of the hydrophobic box around Met-105. It is in direct contact with Tyr-23 and distantly (3.89 Å) with the indole ring of Trp-108.

The side-chain of Trp-111 forms the opposite face of the hydrophobic box and lies between the side-chains of Asn-106 and Lys-116, with the edge of the six-membered ring of the surface of the molecule.

TABLE 5

Relative exposures of tryptophans and half-lives (in s) for exchange of their indole N-H protons

Trp	Relative exposure	Lysozyme	Lysozyme + 0.6M-GlcNAc
62	0.54	<50	(100)
123	0.29	185	180
63	0.19	<50	250
108	0.05	660	7200
111	0.05	1080	2790
28	0.00	v. long	v. long

The indole of Trp-123 lies roughly parallel to the surface of the molecule between residues Ala-122 and Lys-33, partially shielding Cys-30 from contact with surrounding liquid.

(2) Tryptophan-62: Trp-62 lies in the top of the active-site cleft (Figs. 2 and 7) but the only contacts shorter than 4 Å between its indole ring and other residues that are clearly established in the crystal structure are between $N\varepsilon_1$ and the $C\beta$ (3.94 Å) and $C\gamma$ (3.99 Å) of Arg-61. However, neither Trp-62 nor Arg-61 is well located — as shown by the high *B* values illustrated in Fig. 3 — and the side-chain of Arg-73, which is also within possible contact distance, is not well located either. The electron density observed for this well exposed indole ring is consistent with it flapping about the $C\beta$–$C\gamma$ bond and there is evidence for this motion being hindered when inhibitors are bound.

The indole of Trp-62 is also potentially in contact with the side-chain of Gln-121 in a symmetry-related molecule in the crystal structure.

(3) Tryptophan-63: The indole of Trp-63 is buried behind Trp-62 at the top of the active-site cleft (Figs. 2 and 7) between the side-chains of Ile-98 (closest approach 3.51 Å) and Leu-75 (3.51 Å). The $N\varepsilon_1$ atom points out into the cleft and appears, at the present stage of analysis, to be in contact with a water molecule.

(4) Tryptophan-108: Trp-108 (Figs. 2 and 7) is the most buried of the active-site tryptophan groups. It contributes to the surface of the central section of the active-site cleft and forms one face (rather inclined) of the 'hydrophobic box' inside the molecule. The $N\varepsilon_1$, atom is in good-hydrogen-bond contact (2.65 Å) with the carbonyl oxygen atom of Leu-56 and the indole ring makes many contacts shorter than 4 Å with surrounding, mainly non-polarr, groups. In the surface of the cleft $C\beta$ and $C\delta_1$ are also in contact with one of the carboxy oxygen atoms of Glu-35 (distances 3.70 and 3.45 Å, respectively).

N.m.r. evidence. From n.m.r. studies of the tryptophans we have now iden-

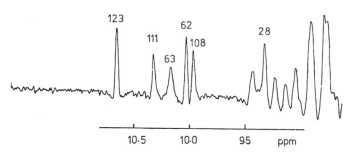

FIG. 8. The six N-H resonances of the tryptophans of lysozyme. (The C-H resonances are given in Fig. 6.)

tified and unambiguously assigned all the six N-H and six C-H protons of the indole rings and some of the 24 protons of the benzenoid rings (see Figs. 6 and 8). We have sufficient data to follow the changes of the six tryptophans individually in any reaction. The resonance frequencies are greatly perturbed from the free amino-acid positions but the general sense of the observed shifts of the resonances is not easily accounted for by reference to the crystal structure. Some resonances are much broadened and we conclude that in the region of Trp-62 and/or Trp-63 there is local motion. The n.m.r. spectra of the tryptophans allow a clear indication that the next-nearest neighbours to the tryptophans in the solution structure are those given by the crystal structure and, to the degree of approximation at which we work, they appear unchanged in position. We shall, therefore, describe the reactivity of the tryptophans by reference to the crystal structure, taking into account mobility as far as possible and following reactions during both n.m.r. spectra and X-ray structure determination of the products.

Tryptophan reactivities. Although pK_a measurements cannot be used to study tryptophan reactions we are able to follow by n.m.r. spectroscopy the exchange rates of ^2H for ^1H for the individual N-H groups of their indole rings. The data are given in Table 5. There is a good correlation between the N-H exchange rates and the exposure factors (Fig. 3). Thus structure does govern reactivity, as shown by this kinetic parameter. Tryptophan-62 is the most accessible tryptophan and turns out to be chemically the most reactive. It is also in a mobile region of the protein.

(1) *Tryptophan-62:* Trp-62 is readily modified by oxidation with *N*-bromosuccinimide to give an indole oxide. The fact that this reaction is fastest for one of the six indoles implies that all the other indoles are more protected from external reagents (Table 5).

The attack causes a small but measurable change in the groove (Table 6).

TABLE 6

Effect of modification of tryptophan-62 on n.m.r. signals

Residue affected	Resonance observed
Trp-63	Indole C(2)-H
	N(1)-H
	C(5/6)-H
Ile-98	γ-CH$_3$
Trp-108	Indole C(2)-H
	N(1)-H

The shifts in the n.m.r. spectrum are important for they cannot arise solely from ring-current changes due to change in Trp-62 and therefore imply that a conformation change must have occurred. The broadening of some resonances indicates that, additionally, some change in motion takes place. The N-H resonance of Trp-63 is also so broadened that it is not detectable and other changes extend to Trp-108. A current study of a crystal of the derivative by X-ray diffraction methods shows a decided movement of the modified Trp-62 itself so that it lies across the groove. In this observed conformation the Trp-62 position would prevent substrate binding. The modified enzyme is almost inactive but it can still bind inhibitors. We have no explanation for this observation as yet. Di- and tri-saccharides bind apparently in an unaltered manner, though more weakly.

Tryptophan-108: Solution studies (Hartdegen & Rupley 1967) showed that treatment with iodine inactivates the enzyme through oxidation of a single tryptophan. Blake (1967) and Beddell *et al.* (1975) showed by experiments on the crystals that treatment with iodine (in KI) leads to a small structural change. They observed a loss of electron density at the position occupied in the native structure by the carboxy group of Glu-35 and an equivalent increase in electron density within covalent bond distance of the Cδ_1 atom of Trp-108. These observations indicated clearly the formation of the oxindolyl ester of Glu-35 which is achieved by a rotation of Glu-35 through 120° about the Cβ–Cγ bond and a small movement of Trp-108. It is, thus, juxtaposition of Trp-108 and Glu-35 that leads to the reactivity of Trp-108 towards iodine. No change in conformation apart from those of Glu-35 and Trp-108 was observed in the crystal structure analysis (Beddell *et al.* 1975). Changes in conformation after this reaction can be followed by n.m.r. spectroscopy of the solution by examining the effects on residues on other parts of the protein (Table 7). The perturbations extend as far as Tyr-23,

TABLE 7

The effect of modification of tryptophan-108 and glutamic acid-35 on n.m.r. signals

Residue affected	Resonance observed
Met-105	S-CH$_3$
Trp-111	C(2)-H N(1)-H
Trp-28	C(2)-H N(1)-H
Ile-98	γ-CH$_3$
Leu-17	C-(CH$_3$)$_2$
Tyr-23	*ortho*-H

on the right hand top surface of the protein and to Leu-17 which is not far from His-15, towards the back of the molecule. There is a clear suggestion here that the internal cross-link produced between Glu-35 and Trp-108 has associated with it redistribution of energy deep into the protein and perhaps generally. Although in itself this chemical step has nothing to do with the mechanism of action of lysozyme it suggests that chemical modifications in proteins can have wide-spread effects which undoubtedly account for some of the confusion of interpretations of chemical probe reactions. However, the conformational changes spread in specific directions, here through the hydrophobic box.

Although this is an artificial (non-enzymic) reaction, it illustrates the general idea of a transfer of information within protein structures. The distance of transference is sufficient for allosteric functioning if there were an allosteric agent that bound close to Leu-17.

Carboxylic acids

The hen egg-white lysozyme molecule incorporates 10 carboxylic acid groups (seven aspartic acids, two glutamine acids and one α-carboxy group). Two of them, Glu-35 and Asp-52 (see Figs. 7 & 9), are essential to the enzyme activity. Before concentrating our attention on these residues and briefly discussing one other in the general vicinity of the active site, we note in passing that one very low pK_a value has been assigned to Asp-66, the most buried of the carboxylic acid groups (Imoyo *et al.* 1972). One of the main challenges that confronts us in understanding the enzymic properties of lysozyme is to learn

TABLE 8

Relative exposures and pK_a values of Asp and Glu residues

Residue	Relative exposure	pK_a	Reference
Asp-18	0.39		Nakae et al. (1975)
Asp-48	0.50	4.4	Kuramitsu et al. (1977)
Asp-52	0.21	3.4	
Asp-66	0.02	1.3	Nakae et al. (1975)
Asp-87	0.75	2.7–4.7	Imoto et al. (1972, p. 665)
Asp-101	0.49	4.4	Nakae et al. (1975)
Asp-119	0.66		
Glu-7	0.50	2.7–4.7	Imoto et al. (1972, p. 665)
Glu-35	0.14	6.1	Kuramitsu et al. (1977)

how best to predict the chemical properties (e.g. the ionization behaviour) of these groups from their environments.

The relative exposures of the aspartyl and glutamyl residues are shown in Fig. 3 and Table 8 together with their pK_a values determined by chemical methods at ionic strength of about 0.1.

The pK_a values of several of these groups (Glu-7, Glu-35, Asp-52, Asp-87 and Asp-101) have been directly confirmed by n.m.r. studies. There is no correlation between pK_a and exposure and we must, therefore, examine the specific chemical environment of each group. Several of the acids with normal pK_a values will be largely ignored since their environment does not affect the pK_a: these are the exposed residues 48, 87 and 7. Asp-52 and Asp-101 have pK_a values near to normal but are in the groove and we shall refer to them in more detail below. The pK_a value of Asp-18 is not yet known with any degree of certainty.

(2) Aspartic acid-101: Asp-101 is well exposed on the surface of the molecule at the top of the active-site cleft. The side-chain is not well defined in the crystal structure (cf. Table 1) but the carboxy oxygen atoms do not appear to be within 4 Å of any other groups in the protein. Some five water molecules have been provisionally located within 4 Å. This acid group is important in substrate binding.

(3) Aspartic acid-52: Asp-52 (Fig. 9a), one of the groups implicated in the catalytic mechanism (Blake et al. 1967b), is located in the left-hand side of the active-site cleft where it has a generally polar environment (Figs. 2 and 7). One of the carboxy oxygen atoms is hydrogen-bonded with the N-δ atom of Asn-59 (2.74 Å away) and is in contact also with the amide groups of Asn-46 (3.60 Å), the carboxy oxygen atoms of Thr-51 (3.87 Å) and Gln-57 (3.82 Å), the main-chain N atom of Asn-59 (3.78 Å) and probably with three water molecules. The other carboxy oxygen atom is in contact with the amide of Asn-46 (3.39

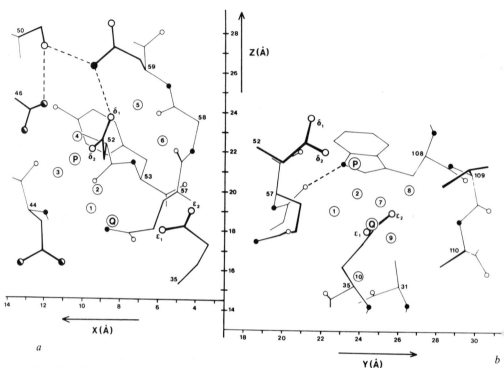

FIG. 9. (*a*) Projection along the *y*-axis of the crystal lattice of atomic positions surrounding Asp-52 and (*b*) projection along the *x*-axis of atomic positions surrounding Glu-35. Probable water molecules are numbered 1–10 and the sites of gadolinium binding are labelled P and Q.

Å) and with 4–6 water molecules or other ligands. This environment suggests a somewhat lowered pK_a

(4) *Glutamic acid-35:* Glu-35 is the second carboxylic acid implicated in the activity of the enzyme and has an environment markedly different from that of Asp-52 (Fig. 9*b*). In the conformation observed in crystals of the native enzyme, one of the carboxy oxygen atoms is in somewhat distant contact with the main-chain nitrogen atoms of Val-109 (3.81 Å) and Ala-110 (3.53 Å) and probably with two water molecules but is otherwise surrounded by non-polar groups: Cδ_1 of Trp-108 (3.53 Å), Cβ of Trp-108 (3.70 Å), Cα of Trp-108 (3.87 Å), Cβ of Ala-110 (3.68 Å), Cβ of Trp-108 (3.70 Å), Cα of Trp-108 (3.87 Å), Cβ of Ala-110 (3.68 Å). The second carboxy oxygen atom is in contact with the Cβ and Cγ atoms of Gln-57 (3.63 and 3.57 Å) and with some six water molecules or other ligands, two of which are also within 4 Å of Asp-52.

N.m.r. studies of the acid groups. The n.m.r. studies of the above groups confirm the conformational positions of some of them, residues 7, 35, 52, 87

TABLE 9

Metal binding to hen egg-white lysozyme

Site	Metal studied	Comment
Asp-101	La, UO_2	Normal reaction for a single carboxylate, $K(La) = 10$
Glu-7	La	Normal $K(La) = 10$
Asp-87	La	Normal $K(La) = 10$
Asp-52 } Glu-35 }	Gd, UO_2	Curious reaction with lanthanides $K(La) = 1000$

and 101, in a general way through perturbation produced by ionization and through the use of lanthanide probes. The second method gives both assignment and conformational parameters. As with the aromatic residues we have no reason to doubt that the positions of the carboxy groups in the molecule in solution are given to a good approximation by the X-ray crystal structure. The n.m.r. methods give both metal-binding constants (Table 9) and pK_a values. As stated earlier, metal binding and pK_a values are normal (Table 8) except for the Asp-52/Glu-35 pair values. We shall concentrate further comment on the three acidic residues of the active-site groove.

(1) Aspartate-101: The examination by n.m.r. spectroscopy of simple physical chemical properties (e.g. pK_a) and the metal binding for Ln(III) (K = 10, Table 9) of Asp-101 show it to be closely similar to a free carboxylate group. The ionization of this group causes only a trivially small perturbation in the n.m.r. spectrum of the protein and we take it that there is no relayed conformational change. In all probability its only function is to form hydrogen bonds to substrates and so to assist binding. If we use this group as a standard for the behaviour of the other RCOOH groups in the groove of the protein then direct comparison with Asp-52 is revealing.

(2) Aspartic acid-52: The estimated pK_a of this group is around 3.0 to 3.5. Its ionization causes a small perturbation of Trp-108. The ionization does not have a marked effect on any other residues of the pocket (see Fig. 7). Despite its significance in the enzyme reaction Asp-52 behaves like a typical surface acid side-chain such as Asp-101. Metal binding in between this anion and glumate-35 is difficult to understand, however (see below).

Glutamic acid-35: The strikingly unusual properties of Glu-35 are shown first by its high pK_a, 6.5. The reaction (1) must be such that the RCOOH form is abnormally stabilized or that the RCOO⁻ group is destabilized. The acid

$$RCOOH \rightleftharpoons RCOO^- + H^+ \qquad (1)$$

form of the enzyme in the crystals shows that the COOH group lies close to the indole ring of tryptophan-108, as shown in Figs. 7 and 9. The group also lies in a 'hydrophobic pocket' which would be expected to cause a high pK_a (Imoto *et al.* 1972). The n.m.r. study of the ionization in solution shows that the inter- action between this residue and Trp-108 is more specific, for a singularly large shift is observed on the indole ring-proton resonances. On titration through the pK_a there are unusual changes in the n.m.r. signals of trypto- phan-108, indicating that some interaction between the two residues has been broken on ionization and the resonances of Trp-108 become more normal. The strong implication is that these groups move away from one another. Thus the specific environment has probably altered the energetics of both the species involved in the ionization, i.e. the ground states of both ·COOH and ·COO⁻ in the groove. The environment must also control the dynamics of the on/off rate of the proton but this is still fast on the n.m.r. time scale, i.e. $> 10^4$ s^{-1}, as a single n.m.r. peak is seen throughout the titration. This indicates that the separation of the two groups Glu-35 and Trp-108, and thus their motions, are faster than substrate conversion. The effect of ionization of Glu-35 is seen in the n.m.r. spectrum on several other residues in the vicinity and there is, therefore, a (small) cooperative effect. A detailed crystallographic study of lysozyme at pH 7 was initiated in Oxford by Dr K. Hamaguchi in 1973 and is now being intensified as the refinement of the structure at pH 4.7 nears com- pletion.

Kuramitsu *et al.* (1977) have shown that when Asp-52 is esterified the pK_a of Glu-35 drops to about 5.25 at ionic strength 0.1 and they take this to be the microscopic pK_a value of Glu-35 when Asp-52 is protonated. The microscopic pK_a of Asp-52 when Glu-35 is ionized is calculated to be about 4.30 and the macroscopic pK_a values for the two groups correspond closely to the microscopic values with Asp-52 ionized and Glu-35 protonated. An interac- tion on ionization of Asp-52 with Trp-108 is seen since the resonance of Trp-108 reflects the state of Asp-52 (see Fig. 10).

THE ASP-52–GLU-35 SYSTEM: A SPECIAL SITE OF METAL BINDING

The description of this site with crystallographic data is based on the structure at pH 4.7, that is below the pH for the ionization of Glu-35. The closest approach between these two carboxy groups is 5.88 Å between Asp-52, $O\delta_2$, and Glu-35, $O\varepsilon_1$, and it is not surprising, therefore, that their ionization properties are coupled to some extent. The electron density includes features between Glu-35 and Asp-52 that have been represented at the present stage of refinement by the water molecules numbered 1 and 2 in Fig. 9. They are,

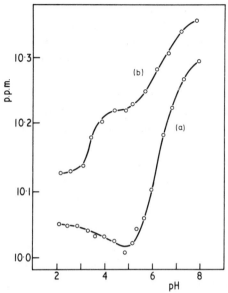

FIG. 10. The titration of Trp-108 N-H resonance in the (a) presence and (b) absence of an inhibitor. The pK_a values of Asp-52 and Glu-35 can be determined in both conditions and are slightly changed by the binding of the inhibitor.

respectively, 3.0 and 2.7 Å from Oδ_2 of Asp-52 and equidistant (4.0 Å) from Oε_1 of Glu-35. However, these two 'water' positions are only 1.51 Å apart and the two sites cannot, therefore, be fully occupied at the same time by water molecules. Each site has an apparent occupancy appropriate to a water molecule and, although the uncertainties at this stage of refinement are large, it seems possible that some other ligand(s) may be involved. Further work, preferably at higher resolution than 2 Å, is needed to characterize this important part of the crystal structure with more certainty.

In the initial determination of the crystal structure by multiple isomorphous replacement (Blake *et al.* 1965) this region was important also for the binding of several of the heavy atoms used in the analysis. Uranyl nitrate and uranyl fluoride, or ligands derived from them, bound at three sites near Asp-52 and Glu-35. Transition metals also bind in this region (Kurachi *et al.* 1975; Teichberg *et al.* 1974) and, more importantly, both n.m.r. and diffraction studies show that lanthanides also bind here (Imoto *et al.* 1972; Morallee *et al.* 1970). They can, therefore, be used in n.m.r. studies as probes of the enzyme's structure and properties (Campbell *et al.* 1975) and consequently we studied the binding of gadolinium(III) in detail crystallographically and by n.m.r. The crystals were prepared by diffusing the gadolinium ion into the lysozyme crystals at pH 4.7. At the present stage of refinement crystallographically

(Perkins 1977; Perkins *et al.* 1978) we see two major peaks near Asp-52 and Glu-35. The overall impression is of an hour-glass-shaped electron density, about 6 Å in length with a diameter of 4 Å at each end and a waist of 2 Å diameter (Fig. 9). The occupancy of site P near Asp-52 is 0.38 and that of site Q close to Glu-35 is 0.23.

Kurachi *et al.* (1975) have studied the binding of Gd(III) to hen egg-white lysozyme crystallographically in the triclinic crystals of lysozyme nitrate. They also observe two sites of binding in similar positions to those described here but with much lower occupancies. In their crystals the site nearer Glu-35 had a higher occupancy (0.03_6) than that nearer Asp-52 (0.01_6).

The binding of gadolinium to lysozyme in the tetragonal crystal form may be influenced by the side-chain of Arg-21 from a symmetry-related molecule which comes within 6 Å of the two gadolinium sites. Co-crystallization of lysozyme with gadolinium results in crystals which, although tetragonal and space group $P4_32_12$, have different cell dimensions: for native lysozyme, $a = b = 79.1$ Å, $c = 37.9$ Å; for lysozyme co-crystallized with Gd, $a = b = 80.8$ Å, $c = 35.4$ Å (Perkins 1977). Structural studies on these crystals are in progress.

N.m.r. studies on metal binding to Glu-35

The n.m.r. data for metal binding to lysozyme in solution have been fully described (Dobson & Williams 1977) and can readily be explained by a single metal-binding site between the two anions. There is reason to suppose that this is not an oscillating binding at the above two distinct sites in rapid equilibrium, since (1) binding is independent of pH and temperature; (2) binding is independent of which lanthanide ion is used (although the ions differ in radius from 0.85 to 1.10 Å; (3) the binding constant is that for two ionized carboxylate groups (see Table 9); and (4) the effect of binding on the Trp-108 resonances shows that the association of the tryptophan with Glu-35 is broken and that Glu-35 is always ionized in the metal-bound form. However, these observations do not eliminate the possibility that there are two binding sites in solution. Further n.m.r. studies are now under way. This study indicates how much care is needed in defining binding sites when there is some mobility near the site.

Associated conformation changes

A preliminary analysis (Perkins *et al.* 1978) of the conformational changes in the crystalline enzyme when Gd(III) is bound has been made by the method of real-space refinement (Diamond 1971). The root-mean-square movement of all atoms with respect to the RS5D lysozyme coordinates (Diamond 1974) was 0.48 Å. These movements include shifts in some of the α-helices and in the β-sheet region, including a significant movement of Tyr-53. At the active site small displacements are observed for Asp-52, Glu-35 and the tryptophans-62, -63 and -108, as a result of which the hydrogen-bond linking Asp-52 with Asn-59 is lengthened from 2.7 to 3.0 Å. Asp-52 and Glu-35 move closer together by about 0.3 Å and the separation of Glu-35 ($O\varepsilon_2$) and Trp-108 ($C\delta_1$) is decreased by 0.2 Å. In contrast to the binding of sugars alone (discussed below) movement of the loop 70–75 is not seen on formation of the Gd–lysozyme or the Gd–N-acetylglucosamine–lysozyme complexes and the movement of Trp-63, although present in the Gd–N-acetylglucosamine–lysozyme complex, is less obvious than in the N-acetyl-β-methyl-D-glucosaminide–lysozyme complex. When these conformational changes are taken into account, site P of Gd(III) binding (Fig. 9) is found to be about 2.6 Å from the $O\delta_2$ atom of Asp-52 and site Q is about 2.4 Å from the $O\varepsilon_1$ atom of Glu-35. The two sites are 3.5 Å apart and cannot be occupied at the same time.

N.m.r. methods also show conformational changes on binding of lanthanides. These include a general tightening of the whole lysozyme molecule, much as is seen on the binding of inhibitors (see below), but the major change is associated with the reorganization around Trp-108. In contrast with the observations in crystals the resonances of Tyr-53 and Thr-51 show no change on La(III) binding.

OTHER GROUPS IN THE GROOVE

Apart from the residues which show reactions there are several hydrophilic and hydrophobic side-chains that are not open to chemical attack. Nevertheless, these residues may play a part in binding and in the control of reactivity. As far as binding is concerned, they provide a stereoselective surface. As far as chemical reactions are involved the motions of many side-chains may be involved in the chemical transformations associated with the reactive path. We note particularly that both n.m.r. and X-ray structural studies show that Val-109 is not in a fixed orientation (see Fig. 7). Often chemical reactions in free solution require solvent rearrangement. In enzyme-catalysed reactions the surface of the enzyme groove is akin to the solvent and its mobility may be required to remove steric strains which would otherwise impede reaction and reduce catalytic efficiency.

SUMMARY OF PHYSICAL AND CHEMICAL PROPERTIES OF LYSOZYME

We have described the physics and the chemistry of the lysozyme molecule. We now consider the binding of inhibitors and substrates to the protein. Before doing so we stress the new features of the enzyme which have been discovered by physical methods and chemical inspection:

(i) the protein has a well maintained fold in all conditions;

(ii) there is mobility in various parts of the protein — particularly side-chain mobility, but also some general vibrations of the main chain;

(iii) there is unconventional chemistry in the active site which is not just due to hydrophobicity but is peculiar to the juxtaposition of particular residues. This contrasts with the rather orthodox chemistry of the surface and the absence of chemistry of the interior;

(iv) there are particular conformational changes associated with some chemical reactions which run some distance from the site of attack;

(v) there is a general conformational change of the whole protein associated with changes of temperature and the binding of some reagents which cross-link the groove. The protein can act as a cooperative unit for energy in these reactions.

INHIBITOR AND SUBSTRATE BINDING

The kinetics of enzyme (E) reactions with substrates (S) are usually analysed in three types of step (reaction 2).

$$E + S \rightleftharpoons ES \dashrightarrow ES^{\neq} \dashrightarrow ES_1 \longrightarrow products \tag{2}$$

There are binding steps, which are the only steps for an inhibitor, and there are steps through transition states, $ES^{\neq}$, to intermediates, ES_1. There can be a multiplicity of such steps and we want to follow the course of them all. Binding in the case of lysozyme has been frequently analysed both by n.m.r. and by X-ray crystallographic study of molecules bound to pre-crystallized enzyme. Here we summarize our observations.

Binding of mono-, di- and tri-saccharides

There are clear indications that the conformation of the enzyme changes in the crystals when inhibitors are bound. The changes extend through the molecule and are most evident in the movement of the loop between residues 70–75, in the clearer definition and reduced motion of Trp-62, and in the angular movement (7°) of the mean-square plane of Trp-63 relative to its

TABLE 10

Effects of inhibitor binding on N(1)-H protons[a]

Inhibitors	Trp-111	Trp-62	Trp-63	Trp-108	Trp-123	Trp-28
GlcNAc[b]	Shifts (−0.08)	Shifts (−0.06)	Shifts (+0.30)	Shifts (−0.24)	0	0
(GlcNAc)$_2$	Shifts (−0.10)	Shifts (−0.10)	Shifts (+0.56)	Shifts (−0.32)	0	Shifts (+0.13)
(GlcNAc)$_3$	Shifts (−0.11)	Shifts (−0.20)	Broadens	Broadens	0	Shifts (+0.15)

[a] Values are in p.p.m. for fully-bound shifts relative to a standard (pH 5.3, $T = 54°C$).
[b] N-Acetylglucosamine. + indicates upfield.

position in native lysozyme. The changes seen in solution are similar. The resonances of Trp-63 show that the region of the protein around Trp-62 and Trp-63 now has a fixed conformation (i.e. there is no longer flapping of a tryptophan). Again the n.m.r. data show a change in the shift of a resonance of Trp-63 and this change is associated with a rearrangement of the binding site *after* initial binding (reaction 3).

$$E + I \rightleftharpoons_{\text{fast}} EI \rightleftharpoons_{\text{slow}} EI^* \qquad (3)$$

The binding of inhibitors also alters the magnitude of the shifts on Trp-108 owing to the Glu-35. As Trp-108 senses strongly the ionizations of both Asp-52 and Glu-35 (Fig. 10), the inhibitor connects the two sides of the groove in some cooperative way.

In solution the effect of binding inhibitors such as N-acetylglucosamine is to shift the tryptophan N(1)-H resonances in the order 108 > 63 > 111. These shifts indicate that inhibitors alter the geometry of the tryptophans somewhat. The effect increases from GlcNAc to (GlcNAc)$_2$ to (GlcNAc)$_3$ (Table 10). The effects are relayed into the hydrophobic box and we see increasing shifts on Met-105 and Ile-98 on going from the mono- to the di- and tri-saccharides. There is some change even of Leu-17 which is just outside the hydrophobic box. The changes are reminiscent of some of those produced by the chemical modification which gives the Glu-35/Trp-108 ester.

As a final example, inhibitor binding alters the resonance positions and the N-H exchange rates of the six tryptophans. The shift data are given in Table 4. The effects are discriminatory from GlcNAc to (GlcNAc)$_3$. The broadening of some resonances also indicates some slow motions inside the protein. The changes in rates of exchange (Table 4) show that only Trp-123 is not affected. It has been known for some time that the rates of N-H exchange (mainly peptide N-H) in lysozyme are grossly affected by the binding of substrates and inhibitors but the n.m.r. data allow these changes to be accurately specified for the first time.

Binding inhibitors to modified protein

The simplest example is the binding of $(GlcNAc)_3$ to the oxindole form of Trp-62. The binding strength is reduced by a factor of 10^2 compared with the native enzyme but the binding can still be easily studied by n.m.r. On binding similar shifts are seen in both the enzyme and the modified enzyme for residues Ile-98, Trp-28 and Trp-63. However, there are differential shifts of Met-105. This indicates that Trp-108 is differently adjusted by the binding of $(GlcNAc)_3$ to the active enzyme as compared to its binding to the *inactive* modified enzyme.

New aspects of the lysozyme mechanism of action

The two functional groups directly involved in the activity of lysozyme are Glu-35 (a proton donor) and Asp-52 (a negative centre for stabilization of a positive centre in the substrate). The activation energy for attack is lowered by the use of the binding energy of the substrate in the distortion of the sugar ring in the critical site D. It is against this background that we review our new information. It may be useful to use Fig. 7 to follow the discussion and to refer to Banerjee *et al.* (1975) for kinetic observations.

The first change in emphasis lies in the observations on Glu-35. N.m.r. data show clearly that Glu-35 and Trp-108 shift relative to one another on ionization of Glu-35. Thus Glu-35 goes through a cycle during catalysis. The possibilities are (4) and (5); in (5) we have avoided writing ·COO⁻ in Site I. Scheme (4) implies that deprotonation is faster than movement but scheme (5) implies the reverse.

$$
\begin{array}{ccccccccc}
-\text{COOH} & \rightarrow & -\text{COO}^- & \rightarrow & -\text{COO}^- & \rightarrow & -\text{COOH} & \rightarrow & -\text{COOH} \\
\text{Site I} & & \text{Site I} & & \text{Site II} & & \text{Site II} & & \text{Site I}
\end{array}
\tag{4}
$$

$$
\begin{array}{ccccccccc}
-\text{COOH} & \rightarrow & -\text{COOH} & \rightarrow & -\text{COO}^- & \rightarrow & -\text{COOH} & \rightarrow & -\text{COOH} \\
\text{Site I} & & \text{Site II} & & \text{Site II} & & \text{Site II} & & \text{Site I}
\end{array}
\tag{5}
$$

N.m.r. observations indicate that, whichever path is followed, movement and deprotonation are both faster than the overall reaction.

With deprotonation Trp-108 also moves so that the energy states of bound substrate (product) are changed in the catalytic cycle. We might say that the enzyme could help not just to distort the substrate but to pull it apart. The cooperativity extends much further up the groove as the substrate cross-links the groove and connects together side-chains along its length. Contact between protein and substrate is intensified after binding by a contraction of the groove which is seen to adjust much of the protein.

Asp-52 has no strictly-comparable behaviour but in the presence of substrate (inhibitor) ionization of Asp-52 causes a change in the interaction between Glu-35 and Trp-108. Although the pK_a of Asp-52 and of Glu-35 and the Glu-35/Trp-108 interaction are similar in the free enzyme and the substrate-bound enzyme, they are not the same and the interaction of Glu-35 with Trp-108 is weaker in the substrate-bound enzyme than in the substrate-free enzyme.

All the results indicate that the enzyme active site is in a specially energized condition. Thus lysozyme does not just distort the substrate on binding toward the transition state of its reaction. It provides a special acid, Glu-35, which is perhaps held close to its transition state for attack on the substrate. This could be achieved through interaction with a special side-chain Trp-108. The view of catalysis which this analysis implies is one which starts from an examination of *two* components in a bimolecular reaction A + B → AB → AB$^{\neq}$ → products. The function of the protein fold, where A is now the protein, is first to raise the ground-state energy of its own attacking and binding groups so that with minor changes of geometry during the passage along the reaction pathway all intermediates are made lower in energy relative to the ground state than would have been possible without the protein fold. Secondly, the binding step to AB functions in two well recognized ways; in establishing the important interaction with Asp-52 and as a means of distorting the substrate B toward its transition-state geometry. Moreover, binding closes the two sides of the pocket together. In addition it now appears from the n.m.r. evidence that substrate binding may alter the interaction of Glu-35 with the other protein side-chains, notably Trp-108 in such a way that the transition state for the reaction, which requires proton donation to the glycosidic oxygen atom, is aided by the preconditioning of the enzyme for the dynamics of the very reaction which it, with the substrate, must go through. This preconditioning is aided by binding but pre-set by the protein fold. Thirdly, the passage through the transition state is a dynamic act. The fold then allows movement at the active site which conforms to the requirement of the best reaction path in a directional sense. These dynamics include a release of the carboxylate anion from Trp-108 and the motion of residues such as valine-109. The change in this interaction energy that allows release and then re-uptake of a proton is no doubt cooperative with the increasing distortion of the substrate toward the broken-bond configuration. The mutual assistance is associated with the changing positions of not only Trp-108 but other groups in the groove which interact cooperatively with Trp-108, e.g. Trp-63.

If these deductions are correct, the function of the groove is much more subtle than the portrait drawn by artists who base their thinking on small-molecule experiments. Our structural studies are fully consistent with the kinetic analysis of the lysozyme reaction by Banerjee *et al.* (1975) and many of the suggestions made by them are now provided with definite experimental support.

CONCLUSION

The joint work in crystal and solution states shows clearly for the first time that the usually accepted statement that the protein sequence is a sufficient cause of the protein structure (Anfinsen 1973) is a first approximation to the important underlying principle. Anfinsen refers to a 'unique three-dimensional structure' and states the principle that this 'native conformation is determined by the amino acid sequence'. We prefer to state that the protein sequence defines 'the partition function of the protein in a chosen environment'. The new statement is not one about structure itself but one about the energetics of various conformations. The extent of the variation in conformation allowed by a given sequence will vary greatly. Thus cytochrome *c* has a confined partition function, but such proteins as chromogranin A, which are virtually random coils, have a wide variety of states open to them. Lysozyme is a case which lies closer to cytochrome *c* in that the sequence confines the protein to a rather narrow set of conformations. Nevertheless we now believe that the allowed variations of conformation, which are strictly controlled, are of the greatest importance and truly reflect the significance of the sequence and its allowed variation.

References

Much of the work described in this paper is as yet unpublished. We have kept references in this paper to a small number: later papers will give broader acknowledgements. Earlier work on which some of the material is based is to be found in Imoto *et al.* (1972) and Dobson (1977).

ANFINSEN, C B. (1973) Principles that govern the folding of protein chains. *Science (Wash. D. C.)* *181*, 223-230

ARNON, R. & SELA, M. (1969) Antibodies to a unique region in lysozyme provoked by a synthetic antigen conjugate. *Proc. Natl. Acad. Sci. U.S.A. 62*, 163-170

AUE, W. P., BARTHOLDI, E. & ERNST, R. R. (1976) Two dimensional spectroscopy. Application to nuclear magnetic resonance. *J. Chem. Phys. 64*, 2229-2246

BANERJEE, S. K., HOLLER, E., HESS, G. P. & RUPLEY, J. A. (1975) Reaction of *N*-acetylglucosamine oligosaccharides with lysozyme. Temperature, pH, and solvent deuterium isotope equilibrium, steady state, and pre-steady state measurements. *J. Biol. Chem. 250*, 4355-4367

BANYARD, S. H., BLAKE, C. C. F. & SWAN, I. D. A. (1974) in *Lysozyme* (Osserman, E. F. *et al.*, eds.), pp. 71-79, Academic Press, New York

BEDDELL, C. R., BLAKE, C. C. F. & OATLEY, S. J. (1975) An X-ray study of the structure and binding properties of iodine-inactivated lysozyme. *J. Mol. Biol. 97*, 643-654

BLAKE, C. C. F. (1967) A crystallographic study of the oxidation of lysozyme by iodine. *Proc. R. Soc. Lond. B Biol. Sci. 167*, 435-438

BLAKE, C. C. F., KOENIG, D. F., MAIR, G. A., NORTH, A. C. T., PHILLIPS, D. C. & SARMA, V. R. (1965) Structure of hen egg-white lysozyme. A three-dimensional Fourier synthesis at 2 Angström resolution. *Nature (Lond.) 206*, 757-761

BLAKE, C. C. F., MAIR, G. A., NORTH, A. C. T., PHILLIPS, D. C. & SARMA, V. R. (1967a) On the conformation of the hen egg-white lysozyme molecule. *Proc. R. Soc. Lond. B Biol. Sci. 167*, 365-377

BLAKE, C. C. F., JOHNSON, L. N., MAIR, G. A., NORTH, A. C. T., PHILLIPS, D. C. & SARMA, V. R. (1967b) Crystallographic studies of the activity of hen egg-white lysozyme. *Proc. R. Soc. Lond. B Biol. Sci. 167*, 378-388

BOTT, R. & SARMA, V. R. (1976) Crystal structure of turkey egg-white lysozyme: results of the molecular replacement method at 5Å resolution. *J. Mol. Biol. 106*, 1037-1046

CAMPBELL, I. D., DOBSON, C. M. & WILLIAMS, R. J. P. (1975) Nuclear magnetic resonance studies of lysozyme in solution. *Proc. R. Soc. Lond. A345*, 41-59

CAMPBELL, I. D. & DOBSON, C. M. (1978) in *Methods of Biochemical Analysis* (Glick, D., ed.), Academic Press, New York, in press

CANFIELD, R. E. (1963) The amino acid sequence of egg white lysozyme. *J. Biol. Chem. 238*, 2698-2707

CANFIELD, R. E. & LIU, A. K. (1965) The disulphide bonds of egg white lysozyme (nuramidase). *J. Biol. Chem. 240*, 1997-2002

DIAMOND, R. (1971) A real-space refinement procedure for proteins. *Acta Cryst. A27*, 436-452

DIAMOND, R. (1974) Real-space refinement of the structure of the hen egg-white lysozyme. *J. Mol. Biol. 82*, 371-391

DOBSON, C. M. (1977) in *Nuclear Magnetic Resonance in Biology* (Dwek, R. *et al.*, eds.), pp. 63-94, Academic Press, London

DOBSON, C. M. & WILLIAMS, R. J. P. (1977) in *Metal-ligand Interactions in Organic Chemistry and Biochemistry*, Part 1 (Pullman, B. & Goldblum, N., eds.), pp. 255-282, Reidel Publishing Company, Dordrecht, Holland

DOBSON, C. M., CAMPBELL, I. D., RATCLIFFE, G. R. & WILLIAMS, R. J. P. (1978) in press

FERGUSON, S. J., LLOYD, W. J. & RADDA, G. K. (1974) An unusual and reversible chemical modification of soluble beef heart mitochondrial ATPase. *FEBS (Fed. Eur. Biochem. Soc.) Lett. 38*, 234-237

FREEMAN, R. & HILL, H. D. W. (1975) in *Dynamic NMR Spectroscopy* (Jackman, L. & Colton, F. A., eds.), pp. 131-153, Academic Press, New York

HARTDEGEN, F. J. & RUPLEY, J. A. (1967) The oxidation by iodine of tryptophan 108 in lysozyme. *J. Am. Chem. Soc. 89*, 1743-1745

IMOTO, T., JOHNSON, L. N., NORTH, A. C. T., PHILLIPS, D. C. & RUPLEY, J. A. (1972) in *The Enzymes*, 3rd edn., vol. 7 (Boyer, P. D., ed.), pp. 665-868, Academic Press, New York

JOLLÈS, J., JAUREGUI-ADELL, J., BERNIER, I. & JOLLÈS, P. (1963) La structure chimique du lysozyme de blanc d'oeuf de poule: étude détaillée. *Biochim. Biophys. Acta 78*, 668-689

JOLLÈS, J., SCHOENTGEN, F., JOLLÈS, P., PROGER, E. M. & WILSON, A. C. (1976) Amino acid sequence and immunological properties of chachalaca egg white lysozyme. *J. Mol. Evol. 8*, 59-78

KURACHI, K., SIEKER, L. C. & JENSEN, L. H. (1975) Metal ion binding in triclinic lysozyme. *J. Biol. Chem. 250*, 7663-7667

KURAMITSU, S., IKEDA, K. & HAMAGUCHI, K. (1977) Effects of ionic strength and temperature on the ionization of the catalytic groups, Asp 52 and Glu 35, in hen lysozyme. *J. Biochem. (Tokyo) 82*, 585-597

LEE, B. & RICHARDS, F. M. (1971) The interpretation of protein structures: estimation of static accessibility. *J. Mol. Biol. 55*, 379-400

LEVITT, M. (1974) Energy refinement of hen egg-white lysozyme. *J. Mol. Biol. 82*, 393-420

LINDERSTRØM-LANG, K. W. & SCHELLMAN, J. A. (1959) in *The Enzymes,* 2nd edn., vol. 1 (Boyer, P. D. *et al.,* eds.), pp. 443-510, Academic Press, New York

McCAMMON, J. A., GELIN, B. R., KARPLUS, M. & WOLYNES, P. G. (1976) The hinge-bending mode in lysozyme. *Nature (Lond.) 262,* 325-326

MORALLEE, K. G., NIEBOER, E., ROSSOTTI, F. J. C., WILLIAMS, R. J. P., XAVIER, A. V. & DWEK, R. A. (1970) The lanthanide cations as nuclear magnetic resonance probes of biological systems. *J. Chem. Soc. Chem. Commun. 18,* 1132-1134

NAGAYAMA, K., WÜTHRICH, K., BACHMANN, P. & ERNST, R. R. (1977) Two-dimensional J-resolved ^{1}H n.m.r. spectroscopy for studies of biological macromolecules. *Biochem. Biophys. Res. Commun. 78,* 99-105

NAKAE, Y., RYO, E., KURAMITSU, S., IKEDA, K. & HAMAGUCHI, K. (1975) pH dependence of the binding constants of N-acetylglucosamine monomers to hen and turkey egg-white lysozymes. *J. Biochem. (Tokyo) 78,* 589-597

PERKINS, S. J. (1977) D. Phil. Thesis, Oxford University

PERKINS, S. J., JOHNSON, L. N., MACHIN, P. A. & PHILLIPS, D. C. (1978) Crystal structures of hen egg-white lysozyme complexes with Gd(III) and Gd(III)-N-acetyl-D-glucosamine. *Biochem. J.,* in press

PHILLIPS, D. C. (1967) The hen egg-white lysozyme molecule. *Proc. Natl. Acad. Sci. U. S. A. 57,* 484-495

RICHARDS, F. M. (1974) The interpretation of protein structures: total volume, group volume distributions and packing density. *J. Mol. Biol. 82,* 1-14

SHINKA, S., IMANISHI, M., MIYAGAWA, N., AMANO, T., IONYNE, M. & TSUGITA, A. (1967) Chemical studies on antigenic determinants of hen egg white lysozyme. I. *Biken J. 10,* 89-107

SHRAKE, A. & RUPLEY, J. A. (1973) Environment and exposure to solvent of protein atoms. Lysozyme and insulin. *J. Mol. Biol. 79,* 351-371

TEICHBERG, V. I., SHARON, N., MOULT, J., SMILANSKY, A. & YONATH, A. (1974) Binding of bivalent copper ions to aspartic acid residue 52 in hen egg white lysozyme. *J. Mol. Biol. 87,* 357-368

WHITE, T. J., MROSS, G. A., OSSERMAN, E. F. & WILSON, A. C. (1977) Primary structure of rat lysozyme. *Biochemistry 16, 1430-1436*

WILLIAMS, R. J. P. (1977) Flexible drug molecules and dynamic receptors. *Angew. Chem. Int. Ed. Engl. 16,* 766-777

WÜTHRICH, K., WAGNER, G., RICHARZ, R. & DE MARCO, A. (1977) in *NMR in Biology* (Dwek, R. *et al.,* eds.), pp. 51-62, Academic Press, London

Discussion

Bränden: Did you see any evidence for movement of the lobes?

Phillips: Glu-35 and Asp-52 moved a little towards one another so that the hydrogen bond from Asp-52 to Asn above it was stretched from 2.7 to 3.0Å and Glu-35 moved a little way out of its pocket.

Gutfreund: Halford (1975) studied the binding of $(GlcNAc)_3$ to lysozyme both by fluorescence and by proton release. The model that fits his results requires a second-order step, in which the fluorescence of Trp-62 is quenched, followed by a first-order isomerization in which the fluorescence of Trp-108 increases (see 1).

$$E + I \rightleftharpoons EI \rightleftharpoons EI^* \tag{1}$$

The isomerization is accompanied by pK changes which can be assigned to proton release (at pH 4.4) from Asp-101 and proton uptake (at pH 6.3) by Glu-35. The fluorescence is quenched in a more aqueous environment and enhanced in a more non-polar environment. These pK changes agree with the idea of neutral acids being affected in this way by aqueous and non-polar environments.

Phillips: When we studied inhibitor binding in the crystals, we observed that Trp-62, which is mobile in the native crystal structure and not clearly seen, becomes much more firmly localized. The left-hand side of the groove (see Fig. 2) closes on to the right-hand side and, although we have not fully analysed the conformational changes, we know that they improve the contact between a sugar bound in site B and Trp-62: there is good contact between the plane of indole and the 'non-polar face' of the sugar.

Gutfreund: We were surprised that the Trp that is already on the surface in more aqueous surroundings becomes even more 'aqueous' and that the one that is in the non-polar environment becomes even more 'non-polar'.

Phillips: Yes, that is easier to understand for Trp-108 than for Trp-62 since the pocket in which Glu-35 lies against Trp-108 (see Figs. 7 and 9) appears to be completely buried by substrate.

Gutfreund: Binding, of course, gives the overall change, but the pK changes which we observe occur only during the second conformational change.

Williams: N.m.r. spectroscopy also showed the steps you described for the two-step relaxation of the tryptophan (see p. 167).

Vallee: I am very much taken by this apparent paradox, which reminds me of an analogous case we came across some time ago with carboxypeptidase A (Riordan *et al.* 1967). We studied the effect of altering specific tyrosyl residues in carboxypeptidase A on its peptidase and esterase activities. Treatment with an eight-fold molar excess of DHT (5-diazo-1*H*-tetrazole) increases esterase activity by 180% and decreases peptidase activity only slightly. Nitration of the native enzyme with tetranitromethane also increases esterase activity by 180% but almost abolishes peptidase activity. In another experiment, nitration of DHT-carboxypeptidase has no further effect on esterase activity but it markedly decreases peptidase activity. At that time we did not know which particular tyrosyl groups were being affected but we concluded that the results were due to modification of two different tyrosyl residues. Since then Cueni & Riordan (1978) have identified the modified residues: in DHT-carboxypeptidase the residue affected is Tyr-248 and in the nitro-enzyme Tyr-248 is nitrated. However, nitration of the DHT-enzyme attacks Tyr-198. Importantly, Tyr-198 is not available for nitration in the native enzyme. This is the only instance of which I am aware in which the reactivity of a second residue is changed on modification of the first. What you described struck me as analogous. This behaviour can best be explained by assuming that there is rotation of a phenylalanine to allow movement of an arginine to activate Tyr-198. There must be some motion of groups and there must be an alteration in the reactivity of Tyr-198 as a result of modification of Tyr-248 with DHT. We still do not know the role of this residue in catalysis, however.

Arigoni: I was surprised by the faster rate of reaction of iodide with Tyr-23 than with Tyr-20 (the more exposed group). However, it is dangerous to take for granted that a more exposed residue will react faster in any given reaction. We should try to focus attention on those factors that may stabilize the corresponding transition-state.

Williams: I didn't get as far as the transition-state: I was suggesting that, with I_2 in KI, the attacking species is probably I_3^- and before one discusses a transition-state one ought to know whether there are any complexes formed in which the I_3^- binds in a specific place. This binding will bias the observed reaction rates (which do not take into consideration concentration factors given by such equilibria). The model suggests that the amines and the arginine surrounding the tyrosine, which is attacked, encourage the approach of the I_3^-.

Arigoni: I accept your argument about local concentration being responsible for a higher rate of reaction but, further along the reaction coordinate, there may be a positively charged species on the tyrosine. Depending on the factors in the immediate surroundings the transition-state may be stabilized or destabilized (reaction may go the same way or the opposite way). Local concentration is not responsible. We need the facts first to disentangle what is going on!

Williams: From the facts available an organic chemist might be able to explain why a particular *surface* residue has a particular reactivity. Maybe the small changes which it is possible to make on protein *surfaces* in isoenzymes give us the best clues to help us to understand mechanisms at active sites, which I believe will be much harder to appreciate than the more local reactions we are discussing just now.

Lipscomb: Carboxypeptidase A contains only one tyrosine which is completely buried—but that residue is iodinated to the greatest extent by I_3^- in the crystalline state. That behaviour raises questions about just what is happening in the I_3^- reaction. In myoglobin I_3^- binds in the same hydrophobic pocket as xenon binds. The use of I_3^- to detect the exposure of a residue may be suspect.

Arigoni: In the odd transfer of the benznitrofuroxan group from the stable intermediate after the reaction with benznitrofuroxan chloride it was claimed that the group was transferred from the O atom of tyrosine to the nucleophilic

N atom of the adjacent lysine in an S_N2-type reaction. Such a reaction demands a linear transition-state. Is it clear from the model that the NH_2 group can approach the CH_2 of the departing group along the axis of the C—O bond of the alkylated benznitrofuroxan species?

Williams: We cannot use the model in that way. None of our measurements allows us to say that any surface lysine is defined in position. The n.m.r. signals from lysine suggest that the tumbling times are 10 times faster than the rotational times of the protein. All the tyrosyl groups have a flipping motion but how much vibrational movement they have as well we do not know.

Arigoni: But can the NH_2 group get behind the benznitrofuroxan group to give a transition-state with an O $\cdots$ C $\cdots$ N bond?

Williams: The transfer of benznitrofuran is definitely intramolecular in the ATPase. The nearest group to Tyr-23 is Lys-115 in lysozyme and this can probably move sufficiently to give such a linear O $\cdots$ C $\cdots$ N transition-state. But you must understand our limitations.

Phillips: Examination of the model suggests that the side-chains of Lys-115 and Tyr-23 can be arranged to give the required transition-state.

Arigoni: I suggest that there remain some chemical studies to be done to clarify that issue.

Knowles: You seem to be claiming that some kinetic consequence derives from the *mobility* of Glu-35 when it ionizes. But when this happens (in the free enzyme) the catalytic activity disappears. Is the movement of Glu-35 that you see in any way relevant to the catalysis by active enzyme, or is it more analogous to, say, the ionization of Ile-16 in α-chymotrypsin? Here, as the pH is raised, Ile-16 ionizes, the conformation of the enzyme changes, and the catalytic activity disappears (Oppenheimer *et al.* 1966). In other words, is the mobility of Glu-35 (as distinct from its ionization state) a necessary part of the catalytic act by lysozyme?

Williams: Yes. My view is that the carboxy group is held initially in the protonated form away from the site in which it will transfer the proton. The group donates a proton as it moves and, as it moves back, the proton is transferred back again. The carboxy group is screened initially.

Additionally the left-hand part of the protein to which the substrate is bound knows about the right-hand side of the groove. For example, when Asp-52 ionizes, in the presence of $(GlcNAc)_3$ Trp-108 senses the ionization. The ionization of Asp-52 seems to help to undo the interaction of Trp-108 and Glu-35. All the interactions and motions seem to be cooperative.

Knowles: But is it a necessary part of the atomic trajectory in the enzyme–substrate complex going to the transition-state?

Williams: I believe so.

Knowles: But why is it a *necessary* part? Surely the action of protonated Glu-35 as a general acid catalyst does not *require* any substantial motion?

Williams: If it acts as a general acid, the proton leaves the carboxy group. At the same time as that action Trp-108 moves away from its position. Thus the movements of the protein groups may alter the energetics of the transition state. If I wanted to make a mechanical analogy to this I would say that at the same time as the COOH moves to transfer the proton the movement of tryptophan stretches the sugar bonds. The whole action is coordinated. The pathway of the reaction does not involve the stationary location of various groups holding attacking entities (e.g. a proton) but, as the reaction proceeds, all the groups immediately around the bond that is being attacked are cooperating to assist progress along the reaction path (i.e. moving to minimize the energy of the pathway). I cannot explain precisely the advantages from having motions as this complex reaction proceeds, but for elementary reactions I can say what advantages motions bring, for example, in electron transfer. Do organic chemists understand (apart from what are good acids and bases in a general way) what particular motions of groups will assist reaction—i.e. the ways that not only the substrate but the surfaces of the catalyst should move for the lowest activation energy of catalysis?

Knowles: Professor Dunitz has proposed lowest energy paths for the motions of groups in several reactions (Bürgi *et al.* 1973; Rosenfield *et al.* 1977) but, if you include the motions of many atoms at an enzyme's active site, then the answer is clearly no, we don't.

Lipscomb: Does Glu-35 form a hydrogen bond to the nitrogen atom of Trp-108?

Phillips: No; nor to anything else in the enzyme (as seen in the crystal structure).

Cornforth: In the conversion of Trp-108 into a oxindole the C-3 atom of the oxindole becomes chiral. Is the chirality known?

Williams: Such a chiral centre is generated in the oxindole from Trp-62 (2) but the Trp-108 reaction is stopped at the ester (1). We grew crystals of the Trp-62 oxindole compound—(that is the *N*-bromosuccinimide reaction)—but we have not found two forms, as there should be if C-3 were a chiral centre. That is to say we have not been able to separate two forms, although we are aware of the problem.

(1)   (2)

Arigoni: Are the measurements ambiguous? Might there be two forms but you have not detected them?

Williams: Yes. No experiment that I have done has allowed me to detect those two forms.

Arigoni: What signal does that C-3 proton give in the n.m.r. spectrum?

Williams: We have not assigned the proton yet.

Kenner: Formation of the oxindole brings the ring across the cleft. Would not the chirality affect its position in the cleft?

Williams: Yes and it will affect the influence of neighbouring ring currents on its proton. The ring current of Trp-63 influences Trp-62 and in principle we ought to be able to say something about the exact conformation.

Phillips: Our present X-ray analysis will show the chiral nature of the oxindole.

Cornforth: Trp-108 seems to be arranged as an electron donor and hydrogen-bonding would accentuate this tendency. Does Trp-108 itself have a role in the stabilization of an oxonium ion in the enzymic mechanism?

Williams: As far as electron donor–acceptor interactions are concerned that is exactly the way we viewed the interaction between the carboxy group and the Trp-108—i.e. as a charge transfer-type of interaction, which would be broken when the carboxy group acquired a negative charge. In other words, the keto part of the carboxy group is the acceptor and the indole is the donor.

Franks: You mentioned that in the proton transfer from Tyr-20 the effect was not propagated further into the protein. What is the mechanism of such proton transfer? Surely the proton is not just transferred to another group? Is it transferred to water? If there is no propagation further into the protein, does this mean that the water acts as a kind of proton sink? How can such a proton transfer stop at the residue at which you observe it?

Williams: By 'no propagational effect' I meant that I could not see any conformational change of other groups in the protein consequent on the ionization of Tyr-23. That does not mean that there is no further protonation–deprotonation reaction within the protein on ionization of such a group—that is an observation that has no direct connection with our tyrosine pK study.

Franks: Are the water molecules seen in the structure so arranged that the effects of proton transfer are dissipated away from the protein?

Phillips: We can see the positions of about 378 water molecules and some Cl⁻ ions distributed around the surface of the protein. As has been mentioned before in this symposium the water structure appears to be more ordered close to the protein and less so away from the protein. But we have to tie this observation in with Professor Richards' observation that, in channels between protein molecules of the sort we see in lysozyme crystals, the water has the properties of bulk water, in many respects.

Richards: I should point out that accessibility measurements not only have an area component but also, for those compounds that are not isotropic, a directional component. The geometry of the iodination is well known. In our work on ribonuclease we came upon a similar situation: a tyrosyl residue which appears to be right out in the solvent but which is not the first to be iodinated. The explanation at that time was that the wrong part of the molecule was exposed to the reagent. Also with regard to accessibility, one finds in general that, as a chain is folded, the carboxy oxygen atoms (or the amino groups of basic residues) tend to lose about half their accessibility to the solvent even though they are still exposed on the surface in the usual sense. What significance, if any, does this have? In general, carboxy groups in proteins have normal pK values apart from overall electrostatic corrections. Do we then have to consider the area of polar groups in a somewhat different context than the non-polar area? Perhaps if a solvent molecule can approach at all, the group may be considered fully accessible. More area, and thus more potential solvent contact, does not alter, for example, ionization properties. If that were true, we should have to adjust our thinking about the effect of partial burying on pK values of ionizable groups.

Phillips: Neither Asp-52 nor Glu-35 is completely buried; they are in contact with water in the crystal (see Fig. 9). That water is more or less structured in the active site crevice. Would you expect water in such a crevice to have the same properties as water anywhere else?

Richards: I imagine that the water in the crevice would be distinctly, detectably, different, in the sense that the tendency of water to leave the crevice might easily differ from that of bulk water by 4 kJ/mol, just by virtue of the radius of curvature of the crevice, without any regard to other factors. I don't know what that would do to the apparent pK values of the affected groups.

Phillips: That is, in terms of substrate binding, would displacement of water from the crevice be easier than might be expected from consideration of bulk water?

Richards: Yes.

Lipscomb: It also many mean that the local dielectric constants are lower.

Franks: We are now getting into the semantics of the dielectric constant of half a dozen molecules—that is dangerous ground; dielectric constant is strictly a macroscopic property of the bulk substance.

Topping: The progressive substitution of acetic acid with bulky alkyl groups

TABLE 1 (Topping)

Changes in pK_a on substitution of acetic acid with bulky groups (in $MeOH/H_2O$ [1/1 v/v] at 40 °C)

Acid	pK_a
CH₃·COOH	5.55
Bu^tCHMe·COOH	6.24
Et₃C·COOH	6.44
Bu^tCMe₂·COOH	6.71
Bu^tCH₂·CBu^tMe·COOH	7.02

causes a significant progressive decrease in acidity (Hammond & Hogle 1955) (see Table 1).

In water as solvent and with further substitution on the α-carbon atom the effect should be even greater. If, as seems reasonable, the effect derives primarily from a change in exposure of the carboxy group to solvent then such studies might offer the possibility of simulating the behaviour of a carboxy group enfolded within protein.

Richards: Those data indicate that the suggestion I made above is probably wrong.

Phillips: A highly substituted molecule of this kind would resemble a sphere with the two oxygen atoms of the carboxy group in the surface exposed to water but shielded behind them.

Johansen: I should caution against attempts to correlate the apparent degree of exposure to bulk solvents and the pK_as and reactivities of amino acid side-chains. In subtilisin, for example, all the carboxy groups are exposed to the solvent but eight of the 16 groups, however, cannot be chemically modified (Adams 1972). Furthermore, even at pH 10, four of these carboxy groups are not ionized (Ottesen & Ralston 1972). In addition, the three tryptophyl residues are exposed to the solvent but they cannot be chemically modified, except when the enzyme is denatured. Thus there seem to be extensive local effects on the reactivity and pK_a values of amino acid side-chains, which are not directly related to their degree of exposure.

Wüthrich: I should like to raise some questions about the potential of crystallographic methods for characterizing dynamic aspects of protein structures. Let me start with a few comments comparing the information gained from n.m.r. spectroscopy and X-ray analysis. A wide gap in credibility has so far dominated evaluations of results obtained by the two methods, mainly justified by the fact that crystallographers can depend on many more data points than n.m.r. spectroscopists can. Recently, the accessible-information content of high resolution n.m.r. spectra of biopolymers has been

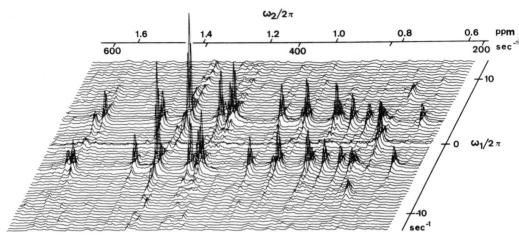

Fig. 1 (Wüthrich). Two-dimensional *J*-resolved 360 MHz ^{1}H n.m.r. spectrum of a 0.01M-solution of the basic pancreatic trypsin inhibitor in D_2O, pD 4.5, $T = 60$ °C. The figure shows the expanded region from 0.5 to 1.7 p.p.m. of a spectrum computed from 64 × 8192 data points. The spectrum contains the resonances of 19 of the 20 methyl groups in the protein. The result is improved resolution (compared with that of the corresponding one-dimensional spectrum) since the multiplet fine structures are no longer manifested on the chemical-shift axis. Rather, as a result of the spread in a second dimension, all lines belonging to a particular multiplet are observed on a straight line which forms an angle of 45° with the ω_2-axis (note the different scales used for the ω_1- and ω_2-axes) and intersects the $\omega_1 = 0$ line at the chemical shift of the proton considered. (Reproduced from Nagayama *et al.* 1977*a*).

greatly increased by the use of high magnetic fields, new pulse techniques and suitable data-manipulation routines (Campbell *et al.* 1973; Wüthrich 1976; De Marco & Wüthrich 1976; Nagayama *et al.* 1977*b*). As an illustration, Fig. 1 shows the high-field region of a *J*-resolved two-dimensional 360 MHz ^{1}H n.m.r. spectrum of the globular protein, basic pancreatic trypsin inhibitor.

An important result obtained from the highly resolved n.m.r. spectra was that the aromatic rings of Phe and Tyr in the interior of globular proteins were not immobilized by their environment but undergo rapid 180° flips about the Cβ–Cγ bond (Wüthrich & Wagner 1975; Wüthrich 1976; Wagner *et al.* 1976). The theoretical model studies I mentioned previously (see p. 42) in which a flexible protein conformation is assumed (Hetzel *et al.* 1976) yielded relations between the rotation state χ^2 of individual aromatic rings and the overall conformational energy of the type shown in Fig. 2. For $\Delta\chi^2 \approx 90°$, the conformational energy was found to be 85–125 kJ/mol (20–30 kcal/mol) higher than for the ring orientation observed in the crystal structure. This corresponds closely to the experimentally determined activation energies for the ring flips (Wagner *et al.* 1976). From Fig. 2 one would also expect that only a small range of χ^2 rotation states near the equilibrium value $\Delta\chi^2 = 0$ is

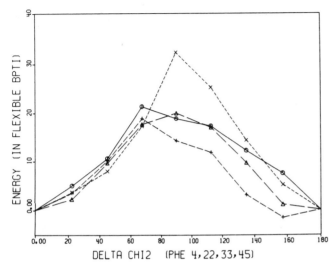

FIG. 2 (Wüthrich). Conformational energies (in kcal/mol) of the basic pancreatic trypsin inhibitor (BPTI) computed for different rotation states χ^2 of the individual phenylalanine rings: o — o, Phe-4; Δ — Δ, Phe-22; +-----+, Phe-33; x····x, Phe-45 (reproduced from Hetzel *et al.* 1976).

populated at ambient temperature. If what is measured as 'temperature factors' in protein does indeed correspond entirely to a temperature factor and is not affected by unresolved disorder, it will manifest essentially the thermally-accessible conformational states rather than large amplitude events which occur with relatively large activation energy, e.g. the flipping motions of aromatic rings. N.m.r. spectroscopy, on the other hand, can give quantitative information on these flipping motions. This particular phenomenon thus presents a nice illustration of the complementary nature of data obtained from X-ray crystallography and n.m.r. spectroscopy.

To cite a specific example, I can add a comment on the data on trypsinogen by Huber's group, to which Professor Blow has already referred. In the X-ray structure part of the polypeptide chain is not visible (Fehlhammer *et al.* 1977). Several years ago, a crystallographer would probably have concluded that the crystal was disordered in that region but now it is concluded that missing diffraction patterns indicate protein mobility. However, on the basis of the X-ray data alone, it appears hardly possible to distinguish between thermal motion and the occurrence of a disordered structure, i.e. a statistical array of different rigid structures which give rise to broadening of the diffraction patterns (personal discussions with Professor R. Huber, 1977). Possibly, these two situations could be distinguished by extending the X-ray studies to low temperatures.

Dunitz: May I add a few words about the general problem of the ability of

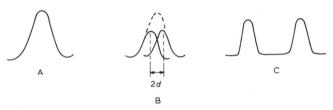

FIG. 1 (Dünitz) Probability distributions of a vibrating atom: A, Gaussian; B, approximately Gaussian superposition of two electron densities in individual unit cells; C, a double maximum for a static disorder.

X-ray analysis to distinguish between disorder on the one hand and thermal motion on the other? Let us consider two extreme situations: in one case the probability distribution of the vibrating atom is approximately a Gaussian function (as would be the case if the potential in which the atom vibrates is almost quadratic) (Fig. 1A). Although I draw this potential in one dimension, one has to remember that the crystallographer, in general, will consider the vibration as a tensor, in other words an equi-probability contour which is represented as an ellipsoid of revolution. To describe such an ellipsoid one needs six numbers: two to define the direction of the major axis of the ellipsoid; one to define the length of the major axis; one for the angle of the second principal axis (which must be in the plane normal to the major axis); one for the length of the second axis; and one for the length of the third principal axis, whose direction is now fixed. (In the tensor description it does not come out that way, but it can always be transformed so as to come out that way.) In addition to x, y, z, the spatial coordinates of an atom, we need a six-parameter temperature factor to describe the anisotropic thermal vibration of that atom. Even if this description works well, to what extent can we distinguish the dynamic motion of the atom from disorder? From X-ray analysis we can observe only the time-averaged and space-averaged electron density. Instead of vibrating, the atom might occupy one position in some unit cells and another position in other cells, in which case the observed electron density will be the superposition of the densities in individual unit cells (Fig. 1B). If the distance between these two positions is not big, the superposition (space-average) density will be practically the same as the Gaussian expected for the vibrating atom (time-average). In the structure-factor calculation the effect of vibration is allowed for by a factor $\exp(-Bs^2)$—a Gaussian function—but for the disorder model the corresponding factor is $\cos(ds)$, where d is half the distance between the two possible sites and $s\,(= 2\sin\theta/\lambda)$ depends on the scattering angle (θ) and the X-ray wavelength (λ). To a first approximation $\exp(-Bs^2)$ expands to $1-Bs^2$ and $\cos(ds)$ expands to $1-d^2s^2/2$. If we set $2B = d^2$, we obtain the same function. The experiment puts a limit on s and, provided *d* is small enough, we cannot distinguish static disorder from the dynamic situation.

In the other extreme we can distinguish because, if the electron density for the static disorder gives a double maximum (Fig. 1C), this is incompatible with a harmonic vibration. In small-molecule crystallography (given good data) it is not too difficult to distinguish one extreme from the other, in cases where it matters—that is to say, where the thermal vibration ellipsoid becomes very anisotropic. Crystallography of proteins is not so fortunate, owing to the limits of resolution: if the two possible disordered positions are closer than, say, 3 Å, one cannot distinguish them in a 3 Å resolution study. I don't know of any serious attempts in protein crystallography to distinguish static from dynamic disorder.

So far I have referred to experiments at one temperature. In principle, one can make measurements at different temperatures. As the temperature is lowered, a genuine vibration ought gradually to get smaller, whereas with disorder several things could happen. One of the alternative sites might become occupied to the exclusion of the other, leading to an ordered structure, or some amount of disorder might be frozen in. However, in general, if cooling does not sharpen the electron density, then disorder rather than thermal vibration would seem to be indicated. One can also do calculations based on some assumed force-field to try to distinguish between the two situations. Some calculations have reproduced the vibrational amplitudes quite well in crystals of small molecules—both simple calculations with Einstein models and more complicated ones with Born—von Karman models.

Professor Wüthrich described the rotation of the benzene ring in terms of a potential-energy profile. But one has to remember that the potential energy is a function in many dimensions; the profile is presumably along the direction of least ascent from the calculated energy minimum, along an eigenvector of the matrix of second derivatives. As Professor Wüthrich said, in principle one can estimate the vibrational amplitude at room temperature from such a profile. That is, if one knows the potential energy, one can guess the amount of thermal vibration. The more quadratic the function is, the better the estimate — if it is quadratic, simple formulae work well — and the less quadratic it is, the more messy the thing becomes. In that case the simple-minded estimate for vibrational amplitudes would be considerably in error.

Lipscomb: In proteins as opposed to small molecules the apparent disorder factors are much larger. When one takes the square root of the *B* values to get something proportional to the mean amplitude one has large numbers. Furthermore, lowering the temperature of a protein crystal (perhaps with an antifreeze agent present) does not greatly sharpen the pattern. This fact argues for considerable disorder and some, but not much, thermal motion.

Dunitz: Professor Phillips reported values of B between 15 and 20 Å.

Phillips: The minimum was about 10 and the maximum was set to be 80, which implies a mean square amplitude of 1 Å.

Dunitz: That means a distance of about 2 Å between atomic sites in a disorder model.

Lipscomb: Typical crystal modes of vibration are about 0.2 Å for most crystals of small molecules, but zero-point vibrational modes in a molecule are 0.01–0.02 Å, except for hydrogen for which they are 0.1 Å.

Phillips: Even in unrefined structures (e.g. the earliest myoglobin maps) one could see groups (e.g. Lys in myoglobin) in two alternative conformations. Several side-chains in insulin have now been observed to have two distinguishable conformations.

Blundell: In the medium-resolution study of insulin we thought that there was an ion pair between a carboxy group (A21) and an arginine (B22). Professor Dorothy Hodgkin now defines the arginine precisely in two positions and even gives the relative occupancies. In some cases we can distinguish between disorder and thermal vibrations. The key point is, as Professor Dunitz said, that we always assume that vibrations are Gaussian. Professor Wüthrich is saying that in some cases they are not and our predictions are wrong. That seems to be the main force of the n.m.r. evidence.

Wüthrich: Even if the different rigid states in a disorder structure were sufficiently far apart so that they could be resolved in the X-ray structure, the polypeptide fragment in question might not be observed, for sensitivity reasons, in cases where there are appreciably more than two different structures, e.g. five or six.

Dunitz: Professor Lipscomb long ago calculated the Fourier Transform of a hindered rotator (King & Lipscomb 1950). Modern computers reduce the work considerably. The question about how meaningful the calculation may be is, of course, another matter.

Roberts: With regard to the difference between two states and many states, it should be remembered that most of the time we cannot tell the difference by n.m.r. spectroscopy either. Most of the processes we are talking about, such as side-chain motion, will be fast on the n.m.r. time-scale, and in these conditions we cannot tell whether the residue is exchanging between 2 states or 22 states, unless—as in crystallography— we can freeze out the separate states.

Lipscomb: Sometimes we can. In some cases of disorder of a protein, backbone or amino acid side-chain we can clearly see two different configurations.

References

ADAMS, K. R. (1972) Investigations on the chemical reactivity of the carboxy groups in subtilisin type Novo. *C. R. Trav. Lab. Carlsberg 38*, 481-498

BÜRGI, H. G., DUNITZ, J. D. & SHEFTER, E. (1973) Geometrical reaction coordinates. II. Nucleophilic addition to a carbonyl group. *J. Am. Chem. Soc. 95*, 5065-5067

CAMPBELL, I. D., DOBSON, C. M., WILLIAMS, R. J. P. & XAVIER, A. V. (1973) Resolution enhancement of protein PMR spectra using difference between a broadened and a normal spectrum. *J. Magn. Resonance 11*, 172-181

CUENI, L. & RIORDAN, J. F. (1978) Functional tyrosyl residues of carboxypeptidase A. The effect of protein structure on the reactivity of Tyr-198. *Biochemistry*, in press

DE MARCO, A. & WÜTHRICH, K. (1976) Digital filtering with a sinusoidal window function—alternative technique for resolution enhancement in FT NMR. *J. Magn. Resonance 24*, 201-204

FEHLHAMMER, H., BODE, W. A. & HUBER, R. (1977) Crystal structure of bovine trypsinogen at 1.8 Å resolution. II. Crystallographic refinement, refined crystal structure and comparison with bovine trypsin. *J. Mol. Biol. 111*, 415-438

HALFORD, S. E. (1975) Stopped-flow fluorescence studies on saccharide binding to lysozyme. *Biochem. J. 149*, 411-422

HAMMOND, G. S. & HOGLE, D. H. (1955) The dissociation of sterically hindered acids. *J. Am. Chem. Soc. 77*, 338-340

HETZEL, R., WÜTHRICH, K., DEISENHOFER, J. & HUBER, R. (1976) Dynamics of aromatic amino acid residues in globular conformation of basic pancreatic trypsin inhibitor. 2. Semiempirical energy calculations. *Biophys. Struct. Mech. 2*, 159-180

KING, M. V. & LIPSCOMB, W. N. (1950) The X-ray scattering from a hindered rotator. *Acta Crystallogr. 3*, 155-158

NAGAYAMA, K., WÜTHRICH, K., BACHMANN, P. & ERNST, R. R. (1977a) Two-dimensional J-resolved ^{1}H n.m.r. spectroscopy for studies of biological macromolecules. *Biochem. Biophys. Res. Commun. 78*, 99-105

NAGAYAMA, K., WÜTHRICH, K., BACHMANN, P. & ERNST, R. R. (1977b) *Naturwissenschaften 64*, 582-583

OPPENHEIMER, H. L., LABOUESSE, B. & HESS, G. P. (1966) Implication of an ionizing group in the control of conformation and activity of chymotrypsin. *J. Biol. Chem. 241*, 2720-2730

OTTESEN, M. & RALSTON, J. (1972) The ionization behaviour of subtilisin type Novo. *C. R. Trav. Lab. Carlsberg 38*, 457-479

RIORDAN, J. F., SOKOLOSKY, M. & VALLEE, B. L. (1967) The functional tyrosyl residues of carboxypeptidase A. Nitration with tetranitromethane. *Biochemistry 6*, 3609-3617

ROSENFIELD, R. E., PARTHASARATHY, R. & DUNITZ, J. D. (1977) Directional preferences of nonbonded contacts with divalent sulfur. *J. Am. Chem. Soc. 99*, 4860-4862

WAGNER, G., DE MARCO, A. & WÜTHRICH, K. (1976) Dynamics of aromatic amino-acid residues in globular conformation of basic trypsin inhibitor. 1. ^{1}H n.m.r. studies. *Biophys. Struct. Mech. 2*, 139-158

WÜTHRICH, K. (1976) *NMR in Biological Research: Peptides and Proteins*, North-Holland, Amsterdam

WÜTHRICH, K. & WAGNER, G. (1975) NMR investigations of dynamics of aromatic amino acid residues in basic pancreatic trypsin-inhibitor. *FEBS (Fed. Eur. Biochem. Sec.) Lett. 50*, 265-269

The free radical in ribonucleotide reductase from *E. coli*

BRITT-MARIE SJÖBERG and ASTRID GRÄSLUND*

*The Medical Nobel Institute, Department of Biochemistry, Karolinska Institute, and *Department of Biophysics, University of Stockholm, Arrhenius Laboratory, Stockholm, Sweden*

Abstract   Protein B2, one of the subunits of ribonucleotide reductase from *Escherichia coli*, contains a stable free radical. It is characterized by a doublet e.p.r. signal centered around $g = 2.0047$ and a sharp peak at 410 nm in the optical spectrum. The radical has been assigned to a tyrosyl residue in the protein with its spin density delocalized over the aromatic ring. Protein B2 also contains two antiferromagnetically-coupled high-spin iron(III) atoms, which stabilize the free radical.

Protein B1, the other subunit of ribonucleotide reductase, contains two binding sites for substrate molecules, which are the four common ribonucleoside diphosphates. It also contains two classes of allosteric effector-binding sites. ATP and deoxyribonucleoside triphosphates function as effectors.

A one-to-one complex of proteins B1 and B2 forms the enzymically-active ribonucleotide reductase. The free radical is, most likely, part of the active site.

The enzyme ribonucleotide reductase catalyses a reaction which is essential to every cell: the formation of deoxyribonucleotides from the corresponding ribonucleotides (Hogenkamp & Sando 1974). This is the exclusive method of producing the precursors for DNA synthesis. In *E. coli* the products of the reaction are the four deoxyribonucleoside diphosphates: dADP, dCDP, dGDP and dUDP. For DNA synthesis the products are further phosphorylated to deoxyribonucleoside triphosphates, although dTTP is synthesized from the dUTP pool by a multistep reaction.

Some of these precursor molecules, as well as ATP, function as (positive or negative) allosteric effectors to modify the activity of ribonucleotide reductase towards its different substrates. Pools of deoxyribonucleotides are usually small and ribonucleotide pools comparatively large. It is, therefore, likely that the reduction of ribonucleotides is not only essential but also one of the rate-controlling steps in DNA synthesis. In *E. coli*, for instance, two genes of the same operon code for ribonucleotide reductase. Conditional mutations in

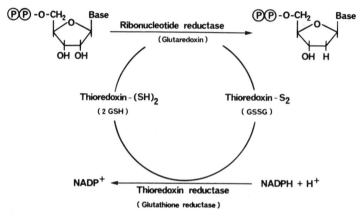

F<small>IG</small>. 1. Components of the *Escherichia coli* ribonucleotide reductase system. The glutathione
hydrogen-donor system is shown in parentheses.

any of the genes immediately give rise during restrictive conditions to defective DNA synthesis and, as a consequence, cell death (Fuchs *et al.* 1972; Fuchs
& Karlström 1973).

Ribonucleotide reductase from *E. coli* consists of two subunits, denoted
proteins B1 and B2. Protein B1, the larger subunit, is composed of two
unidentical but similar polypeptide chains with a total molecular weight of
160 000. The difference seems to reside mainly in the NH_2-terminal part of the
polypeptides. The B1 subunit contains several cysteinyl residues in its active
form. Four of these are part of the active site of ribonucleotide reductase and
it is these thiol groups that are reversibly oxidized to disulphide bridges
during the enzymic reaction (Thelander 1974). *In vitro* studies on separated
subunits have shown that all ribonucleoside diphosphate substrates as well as
nucleoside triphosphate effectors bind to protein B1 alone (von Döbeln &
Reichard 1976). The dissociation constant for a given substrate depends on
the effector molecule present, such that the regulation of the binding of
substrates to the B1 subunit mimics the allosteric regulation of the activity of
ribonucleotide reductase. However, the active form of ribonucleotide reductase is formed in the presence of Mg^{2+} as a one-to-one complex of proteins B1
and B2. Protein B2, the smaller subunit, has a molecular weight of 78 000 and
consists of two identical polypeptide chains. It contributes an organic free
radical, which is directly correlated with enzymic activity (Ehrenberg &
Reichard 1972) and which is stabilized by two iron atoms (Atkin *et al.* 1972).

The electrons needed for the reduction of ribonucleotides are provided by
NADPH and are transported to ribonucleotide reductase by the thioredoxin
system (Fig. 1). This consists of two well-characterized proteins: thioredoxin

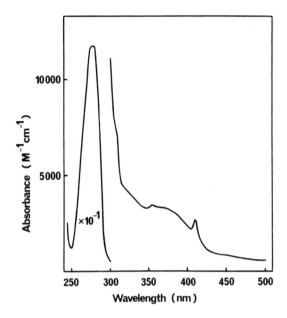

FIG. 2. Light absorption spectrum of protein B2 at +4 °C.

reductase and thioredoxin. Thioredoxin reductase contains redox-active thiol groups and FAD in the active site (Thelander 1970). Thioredoxin, which is a low-molecular-weight protein, alternates between a reduced and an oxidized form through a single pair of redox-active cysteinyl residues (Holmgren *et al.* 1975).

E. coli strains that are deficient in thioredoxin grow as well as normal strains (Mark & Richardson 1976) and in such mutant cells yet another electron-transport system, composed of glutathione, glutathione reductase and a new small protein, called glutaredoxin, can carry electrons from NADPH to ribonucleotide reductase (Fig. 1) (Holmgren 1976). Both transport systems seem to be active *in vivo* in normal *E. coli* cells.

The radical in protein B2 is easily recognized by a sharp peak at 410 nm in the light-absorption spectrum and a characteristic e.p.r. signal centred around $g = 2.0047$ (Figs. 2 and 3). The main feature of the e.p.r. signal is a doublet splitting of about 19 G (1.9 mT). In addition the spectrum shows a partly resolved triplet splitting of about 7 G with the relative line intensities 1:2:1.

What is the molecular structure of the radical in protein B2? The problem was tackled by use of isotopic substitution experiments, which for a long time were completely negative. The signal appeared unchanged in preparations of protein B2 enriched with ^{57}Fe or [^{15}N]histidine, and in generally-^{15}N-enriched

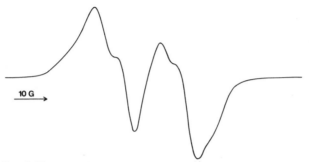

FIG. 3. Electron spin resonance spectrum of protein B2 at 77 K.

samples. Also, substitution of most exchangeable protons for deuterium in B2 resulted in the normal doublet.

Recently a strain of *E. coli* was isolated which overproduces ribonucleotide reductase (O. Karlström, unpublished work) and so enabled us to make a considerably more detailed study of the radical in protein B2. The strain is lysogenic for a defective, heat-inducible hybrid λ bacteriophage, which in its DNA also carries the structural genes for both subunits of ribonucleotide reductase. In special growth conditions ribonucleotide reductase constitutes about 15% of the extractable protein of the cells, a level which is 50 times higher than the levels in normal *E. coli* strains (Eriksson *et al.* 1977). For technical reasons the content of protein B2 is about 1% of the total protein in the experiments reported below. This level is high enough for the protein B2-specific e.p.r. signal to be studied directly in samples of frozen thick suspensions of whole cells. A culture volume of 60–120 ml is usually sufficient to give an e.p.r. signal with reasonable signal-to-noise ratio. For isotope-substitution experiments we added isotopically labelled ingredients to the growth medium (Sjöberg *et al.* 1977). A drastic change in the B2-specific e.p.r. signal was observed when cells were grown in a medium based on 93% deuteriated water instead of normal protonated water: the signal appeared as a singlet instead of the normally observed doublet indicating that the doublet splitting of the spectrum is caused by hyperfine coupling to a hydrogen atom. Inclusion of different protonated amino acids into the deuteriated growth medium more specifically located the radical to a tyrosyl residue.

We determined the site of the unpaired electron within the tyrosyl residue by growing cells in normal medium in the presence of specifically deuteriated tyrosine (Gräslund *et al.* 1978) and observed that [β,β-^{2}H$_2$]tyrosine was responsible for the conversion of the main doublet into a singlet. Addition of [3,5-^{2}H$_2$]tyrosine resulted in a loss of the partly resolved triplet splitting, and addition of [2,6,α-^{2}H$_3$]tyrosine gives a signal similar to that in the normal

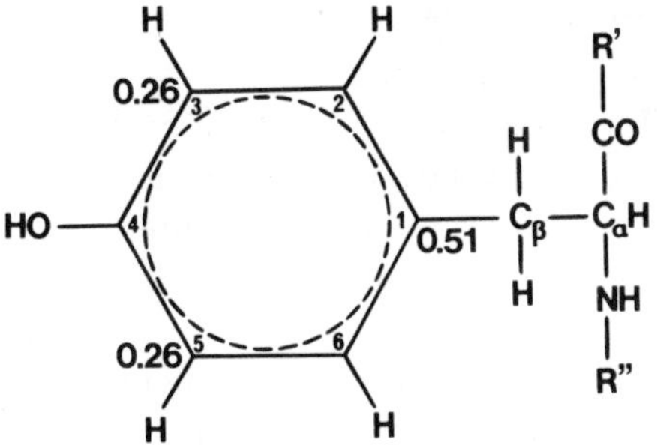

FIG. 4. Structure of the tyrosyl radical in protein B2: relative spin densities on C-1, C-3 and C-5 are indicated.

B2-spectrum. All these facts are consistent with a tyrosyl radical with the unpaired electron spin residing in the aromatic ring and having measurable spin densities on C-1, C-3 and C-5 (Fig. 4). The spin density on C-1 gives rise to an isotropic hyperfine coupling to one of the β-hydrogen atoms through hyperconjugation, which causes the major doublet splitting of the e.p.r. spectrum. The magnitude of the hyperfine-coupling constant should depend strongly on radical geometry (Morton 1964). No rotation around the bond between the aromatic ring and the methylene group can make the two methylene protons geometrically inequivalent. Thus, for a locked steric configuration where one Cβ–H bond is in the plane of the aromatic ring, the hyperfine coupling to that hydrogen atom will be negligible, while the other hydrogen atom is in a steric arrangement favourable for hyperfine coupling through hyperconjugation. Anisotropic hyperfine coupling to H-3 and H-5 arises from the spin densities on C-3 and C-5 and yields the partly resolved triplet splitting.

The interpretation of the radical structure is confirmed by two other growth experiments. Addition of generally-^{13}C-labelled tyrosine to the growth medium results in a broad and structureless e.p.r. signal in agreement with what is expected for a radical with its unpaired electron delocalized in the aromatic ring. Finally, the addition of [β-^{13}C]tyrosine results in a somewhat broadened (<5 G) but otherwise 'normal' e.p.r. signal, indicating that the unpaired electron has considerable spin density on a carbon atom neighbouring the β-carbon atom but not on the β-carbon atom itself. The distribution of spin densities estimated for each position from the observed hyperfine-coupling constants is: C-1, 0.51; C-3, 0.26; and C-5, 0.26. These values are similar to those observed by Box et al. (1974) for one of the products formed on X-ray irradiation of single crystals of tyrosine. This radiation product was tentatively identified as a primitive oxidation product of tyrosine, formed by the loss of an electron from the aromatic ring.

What is the role of iron in protein B2? The enzymic activity of protein B2 depends on the presence of the radical and the radical in turn depends on the presence of iron. Mössbauer spectroscopic studies revealed an antiferromagnetically coupled binuclear Fe(III) complex and two classes of iron in B2 (Atkin et al. 1972). It seems likely that each binuclear complex is intrinsically asymmetrical and contains one iron atom of each class. Radical content has no effect on the distribution between the two iron populations. For instance, treatment of protein B2 with hydroxyurea or hydroxylamine, which destroys the radical, does not affect the iron in any detectable way. The same type of radical loss can also occur in B2 by mere ageing of the enzyme in solution. Aged or inactivated B2 may be reactivated first after removal of the iron to yield a preparation (denoted apoprotein B2) devoid of both radical and iron. When iron is reintroduced into apoprotein B2 in the form of a Fe(II)–ascorbate complex, the radical is regenerated and activity is restored. This implies that the role of iron in B2 is not simply a structural one but rather to generate and stabilize the radical.

With the exception of the 410 nm band the electronic and Mössbauer spectra of protein B2 resemble corresponding spectra of oxy- and met-haemerythrins. The iron atoms in these proteins are antiferromagnetically coupled and in the high-spin Fe(III) state. At least one and sometimes two tyrosyl residues are ligands of the iron complex in this case (Klotz et al. 1976).

Does the radical of protein B2 play a role in the enzyme reaction? Recent experiments with substrate analogues show that the radical in B2 can be scavenged by 2′-deoxy-2′-azidoribonucleoside diphosphates (Thelander et al. 1976). These analogues specifically inactivate the B2 part only when an enzymically-active complex between proteins B1 and B2 is formed; B2 is inactivated even in the presence of catalytic amounts of protein B1, provided that a hydrogen-donor system is present to regenerate the redox-active dithiols in B1. Consequently, the azido derivatives must bind to the active site of the enzyme and be reduced in order to destroy the radical in B2. So far this is the sole evidence that the radical in B2 is part of the active site and participates directly in ribonucleotide reduction.

The maximal radical content usually obtained in recent B2 preparations is 1.0 unpaired spin per B2 molecule or, expressed differently, one radical per two iron atoms (Eriksson et al. 1977). There is no apparent loss of organic radical during the preparation procedure, although this does not exclude the existence of a B2 protein with even higher radical content. If, however, the current figure represents a maximal value, it implies an enzyme with only one

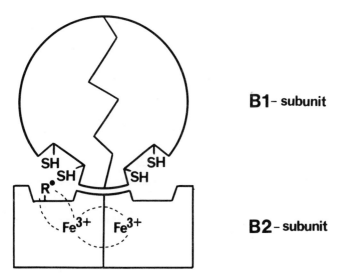

B1- subunit

B2- subunit

FIG. 5. Proposed active site of ribonucleotide reductase.

radical and with two catalytic sites, which are formed from both subunits of ribonucleotide reductase (Fig. 5). Such an organization may imply a reaction mechanism with alternating active sites (Smith & Boyer 1976).

References

ATKIN, C. L., THELANDER, L., REICHARD, P. & LANG, G. (1972) Iron and free radical in ribonucleotide reductase. Exchange of iron and Mössbauer spectroscopy of the protein B2 subunit of the *Escherichia coli* enzyme. *J. Biol. Chem.* 248, 7464-7472

BOX, H. C., BUDZINSKI, E. E. & FREUND, H. G. (1974) Effects of ionizing radiation on tyrosine. *J. Chem. Phys.* 61, 2222-2226

EHRENBERG, A. & REICHARD, P. (1972) Electron spin resonance of the iron-containing protein B2 from ribonucleotide reductase. *J. Biol. Chem.* 247, 3485-3488

ERIKSSON, S., SJÖBERG, B.-M., HAHNE, S. & KARLSTRÖM, O. (1977) Ribonucleoside diphosphate reductase from *Escherichia coli*. An immunological assay and a novel purification from an overproducing strain lysogenic for phage λdnrd. *J. Biol. Chem.* 252, 6132-6138

FUCHS, J. A. & KARLSTRÖM, H. O. (1973) A mutant of *Escherichia coli* defective in ribonucleoside diphosphate reductase. 2. Characterization of the enzymatic defect. *Eur. J. Biochem.* 32, 457-462

FUCHS, J. A., KARLSTRÖM, H. O., WARNER, H. R. & REICHARD, P. (1972) Defective gene product in *dna*F mutant of *Escherichia coli*. *Nat. New Biol.* 238, 69-71

GRÄSLUND, A., EHRENBURG, A., SJÖBERG, B.-M. & REICHARD, P. (1978) *J. Biol. Chem.*, in press

HOGENKAMP, H. P. & SANDO, G. N. (1974) The enzymatic reduction of ribonucleotides. *Struct. Bonding (Berlin) 20,* 23-58

HOLMGREN, A. (1976) Hydrogen donor system for *Escherichia coli* ribonucleoside-diphosphate reductase dependent upon glutathione. *Proc. Natl. Acad. Sci. U.S.A. 73,* 2275-2279

HOLMGREN, A. SÖDERBERG, B.-O., EKLUND, H. & BRÄNDÉN, C. I. (1975) Three-dimensional structure of *Escherichia coli* thioredoxin-S_2 to 2.8 Å resolution. *Proc. Natl. Acad. Sci. U.S.A. 72,* 2305-2309

KLOTZ, I. M., KLIPPENSTEIN, G. L. & HENDRICKSON, W. A. (1976) Hemerythrin: alternative oxygen carrier. *Science (Wash. D. C.) 192,* 335-344

MARK, D. F. & RICHARDSON, C. C. (1976) *Escherichia coli* thioredoxin: a subunit of bacteriophage T7 DNA polymerase. *Proc. Natl. Acad. Sci. U.S.A. 73,* 781-784

MORTON, J. R. (1964) Electron spin resonance spectra of oriented radicals. *Chem. Rev. 64,* 453-471

SJÖBERG, B.-M., REICHARD, P., GRÄSLUND, A. & EHRENBERG, A. (1977) Nature of the free radical in ribonucleotide reductase from *Escherichia coli. J. Biol. Chem. 252,* 536-541

SMITH, O. J. & BOYER, P. D. (1976) Demonstration of a transitory tight binding of ATP and of committed P_i and ADP during ATP synthesis by chloroplasts. *Proc. Natl. Acad. Sci. U.S.A. 73,* 4314-4318

THELANDER, L. (1970) The amino acid sequence of a peptide containing the active center disulfide of thioredoxin reductase from *Escherichia coli. J. Biol. Chem. 245,* 6026-6029

THELANDER, L. (1974) Reaction mechanism of ribonucleoside diphosphate reductase from *Escherichia coli.* Oxidation-reduction-active disulfides in the B1 subunit. *J. Biol. Chem. 249,* 4858-4862

THELANDER, L., LARSSON, B., HOBBS, J. & ECKSTEIN, F. (1976) Active site of ribonucleoside diphosphate reductase from *Escherichia coli.* Inactivation of the enzyme by 2'-substituted ribonucleoside diphosphates. *J. Biol. Chem. 251,* 1398-1405

VON DÖBELN, U. & REICHARD, P. (1976) Binding of substrates to *Escherichia coli* ribonucleotide reductase. *J. Biol. Chem. 251,* 3616-3622

Discussion

Battersby: Is there no unpaired spin on the oxygen atom of the tyrosyl residue? The figures you quoted for the charge distribution added up to 100% on carbon atoms.

Sjöberg: It is because the charge distribution on C-1, C-3 and C-5 add up to unity that we think there is very little spin density on the oxygen atom. An experiment with ^{17}O-tyrosine would be conclusive.

Knowles: We have found recently that both aromatic and aliphatic azides are rapidly reduced by dithiols (J. V. Staros, H. Bayley & D. N. Standring, unpublished work). For instance, dithiothreitol reduces azides to amines about 1000 times faster than, say, cysteine or 2-mercaptoethanol. Is your azido compound (2'-deoxy-2'-azidocytidine 5'-diphosphate) reduced? The enzyme as you described it presumably involves a disulphide element in its catalytic action. Is the azido compound reduced to the amino compound, or does it go all the way to the 2'-deoxy derivative?

Sjöberg: Analyses of the product after inhibition of protein B2 with azido analogues only exclude the possibility that the amino analogue was formed; the deoxy form could not be separated from the azido form in the analyses that were performed. The inhibition of protein B2 is monitored as the disappearance of the 410 nm absorption band; at the same time the redox-active dithiols of the B1 subunit are consumed (Thelander *et al.* 1976).

Knowles: That would fit with the possibility of a facile reduction of the azide to an amine by a *vicinal* dithiol.

Sjöberg: Yes.

Williams: To generate the free radical you added ascorbate and iron(III), and so presumably oxygen as well. The tyrosine is oxidized to give a free radical and there is no coupling between that free radical and the iron at the end of the reaction—i.e. the free radical is separated from the iron—and a high-energy state is made within this protein. The free radical in tyrosine has an oxidation potential of 0.8 V which is just about the same potential as would result from reaction of oxygen with iron at pH 7. Perhaps the iron is needed as a catalyst to make the free radical but has then nothing else to do with the mechanism. In other words its action is like that of an experimentalist making a defect in a solid (a well known way of making catalysts in the solid state). In the B_{12} enzyme series there is another set of enzymes that catalyse this same internal reaction. An initial cobalt–carbon bond is broken to make a free radical which resides on a sugar — that is, another high-energy state (see Williams 1971 for review). Why should reduction of ribose to deoxyribose require such high-energy free radicals?

Arigoni: It is not easy to see why it should, but it does. To remove such a hydroxy group the last method that I would consider is a radical mechanism. Has the corresponding amino compound been investigated as a possible substrate for the reaction?

Sjöberg: 2'-Deoxy-2'-amino-CDP binds to the substrate-binding site of ribonucleotide reductase but it is not reduced (U. von Döbeln, unpublished results, 1975).

Cornforth: If the proton still remains on the tyrosyl hydroxy group, the molecule is either a radical anion or a radical cation. Which is it?

Sjöberg: We favour the cation, but we do not even know if the proton still is on the hydroxy group. It might be that the iron atom complexes that hydroxy group.

Williams: The e.p.r. data rule that out. There is a two-iron centre which contains iron(III) and a normal organic radical which, according to your colleague Dr Ehrenberg, is not bound to the iron atoms.

Sjöberg: Although no hyperfine coupling is seen, the iron is probably close to the radical, since it is needed for the generation of the radical. Box *et al.* (1974) could distinguish three products after X-ray irradiation of single crystals of tyrosine. A primitive oxidation product, which was formed by the loss of one electron, had measurable spin densities on C-1, C-3 and C-5, negligible spin densities on C-2 and C-6 and low spin density on the hydroxy oxygen. Of the other two products, one was an aryloxy radical formed by the loss of one electron and one proton, the other was a reduction product. The coupling parameters of the latter were very different from ours and the major feature of the aryloxy radical was a considerable spin density on the hydroxy oxygen. So our observed distribution of spin densities resembles that of the primitive oxidation product more than that of the other two products.

Lipscomb: That implies the cation, but molecular orbitals can be drawn that would make it look more like the anion.

Wüthrich: Can the possibility that the spin density is delocalized from the iron centre onto the tyrosine ring be excluded?

Williams: Yes. Dr Ehrenberg and his co-workers have established this (see p. 189).

Sjöberg: Since the conjugated ring of tyrosine can stabilize electron vacancies, the free radical could theoretically be generated elsewhere in tyrosine or in the protein and then be transferred to this stable position.

Arigoni: Is the rate of appearance of the e.p.r. signal for the tyrosyl radical compatible, from a kinetic point of view, with the rate of reduction of nucleotide?

Sjöberg: We have not looked at the kinetics of the e.p.r. signal yet. A transient doublet e.p.r. signal is observed in ribonucleotide reductase from *Lactobacillus leichmannii* during the enzymic reaction (Sando *et al.* 1975). It has been studied by use of rapid freeze-quenching techniques. In our case the doublet e.p.r. signal is stable and we have only used ordinary mixing techniques, in which experiments we could not observe any change in the signal during catalysis. We have not yet done any rapid freeze-quenching experiments with our system.

Wüthrich: How does the life time of this radical compare with the life time of the radicals in irradiated tyrosine single crystals?

Sjöberg: The radical in protein B2 is stable as long as the protein persists. There is no indication of loss of the free radical during the isolation of protein B2. In fact, we observe that the total amount of protein B2 specific radical increases 2–5 times during the initial preparative steps to give a purified B2 enzyme containing 1.0 unpaired spin per molecule (Eriksson *et al.* 1977).

Williams: A similar sort of free radical appears in cytochrome *c* peroxidase and is required for electron transfer between two proteins. So at least one other example is known but I have not compared the e.p.r. signals.

References

SANDO, G. N., BLAKLEY, R. L., HOGENKAMP, H. P. C. & HOFFMAN, P. J. (1975) Studies on the mechanism of adenosylcobalamin-dependent ribonucleotide reduction by the use of analogs of the coenzyme. *J. Biol. Chem.* 250, 8774-8779

WILLIAMS, R. J. P. (1971) *Inorg. Chim. Acta 5*, 137-155

Other references are given on pp. 193 and 194.

Dynamics of local conformation and enzyme function

BERT L. VALLEE and JAMES F. RIORDAN

Biophysics Research Laboratory, Department of Biological Chemistry, Harvard Medical School, and the Division of Medical Biology, Peter Bent Brigham Hospital, Boston, Massachusetts

Abstract   Enzyme function is critically affected by gross changes in tertiary structure, but much less is known about the dependence of activity on localized conformational changes that might occur during the catalytic process. Such information is essential to the understanding and verification of enzyme mechanisms, and can only be obtained from systems in which the substrate binding and catalytic groups of the enzyme active centre are not only identified but also rendered appropriate probes of the catalytic reaction, that is they must signal changes in their mutual interactions at rates at least as fast as catalysis. It should then be possible to observe directly each of the dynamic events that result in catalysis. Time-averaged structural analyses can neither reveal the dynamics of catalysis nor describe the conformational details of molecules whose structures—particularly of the active centre—are motile. Moreover, the three-dimensional structures of enzymes in crystals and solutions may not always be identical as evidenced by marked differences between the kinetic properties of crystalline and of dissolved enzymes (e.g. hexokinase, glycogen phosphorylase and carboxypeptidases A and B).

Various experimental approaches have been devised to explore the relationship between structure and function of these and other enzymes. We have focused on the syncatalytic, spatial relationships of active-site residues as they bear on the mechanism of enzyme action using absorption, circular dichroism, magnetocircular dichroic, and resonance Raman spectroscopy as well as resonance energy transfer. Stopped-flow, pH and temperature jump methods quantitatively assess both the conformational and local structural features of the enzyme and reveal multiple, discrete conformational states that prove to have mechanistic significance. These and related data serve as a basis for a minimal model of the dynamic aspects of enzyme action.

Enzymic catalysis is the consequence of a series of molecular, structural events initiated by the acquisition of tertiary structure of the newly-synthesized polypeptide and usually terminated by a conformational relaxation that accompanies dissociation of the enzyme–product complex. Protein chemists and enzymologists, among others, have devised numerous ingenious experimental approaches to explore and elucidate individual aspects of this structure–function relationship. Enzyme denaturation, for example, and the resultant loss of catalytic activity have long been recognized to be due to gross changes in protein structure. In fact, the process of denaturation can only be defined in structural terms and with reference to native conformation.

A host of physical and chemical techniques has been employed to investigate denaturation, most of them being relatively insensitive, yet adequate for the purpose, since the underlying structural changes are so profound. On the other hand, only a limited number of approaches have been devised to examine the subtle conformational changes that occur during the actual catalytic process itself. In general, these require systems in which the substrate-binding and catalytic groups are not only identified but also rendered appropriate probes of the catalytic reaction. If such groups can signal changes in their mutual interactions at rates at least as fast as catalysis, it becomes possible to observe directly each of the accompanying dynamic events. This information is clearly essential to the comprehension and verification of enzyme mechanisms.

Much of our current understanding of the mode of interaction of substrates with the active sites of enzymes derives from X-ray structure analysis of enzyme–substrate or enzyme–inhibitor complexes or both. Indeed, such studies have even served as the basis for deductions regarding the *dynamics* of enzyme action, although the time-averaging nature of the approach precludes the direct observations of transient intermediates. Implicit in these deductions was the hypothesis that the structure of an enzyme in the crystalline state is the same as that in solution. However, numerous investigations have indicated that enzymes in solution have multiple, rapidly-interconverting conformations (Harrison *et al.* 1975; Weber 1975; Blake 1976) as predicted by Linderstrøm-Lang & Schellman (1959). The process of crystallization introduces interactions (e.g. crystal packing forces) with energetics comparable to those necessary to maintain particular conformations themselves. Thus, the three-dimensional structures of enzymes in crystals and solutions need not always be identical. Indeed, different crystal forms could well comprise multiple and/or different populations of enzyme conformers, as has been shown, for example, for different crystal forms of hexokinase which exhibit variable catalytic properties (in fact, one is completely inactive) (Anderson *et al.* 1974). Ideally, any conclusions about the structural basis for activity should rely on determinations of structure and activity using the same material. This would seem imperative if such data were to serve to define a mechanism of action. Catalytic activity of enzymes is generally measured with enzyme solutions, but comprehensive structural analyses cannot yet be performed in that physical state. Hence, determination of the activity of enzyme crystals would seem to be the only feasible alternative approach for such comparisons at this time. This approach might at least indicate whether or not the process of crystallization is associated with changes in activity. Detailed kinetic studies on, for example, carboxypeptidase A (Quiocho & Richards 1966; Spilburg et al. 1974, 1977), carboxypeptidase B (Alter et al. 1977) and glycogen phosphorylase (Kasvinsky & Madsen 1976) have revealed that the predominant effect of crystallization is a large decrease in k_{cat}. However, reductions in k_{cat} need not accompany enzyme crystallization. Bayne & Ottesen (1976) have found that the activity of pig-heart lactate dehydrogenase is not affected by changes in physical state and similar observations have been reported for other systems (Rossi & Bernhard 1970; Sawyer 1972). Thus, no generalization about the effects of crystallization on enzyme function has yet emerged. Nevertheless, since the integration of functional data obtained in solution with the three-dimensional structure derived from crystals remains an important means of discerning the mode of action of enzymes, interpretations of crystal structures, in mechanistic terms, must be firmly based on a detailed examination of the activity associated with the structure in the particular crystal form examined. Fortunately, detailed kinetic analysis of catalysis by crystals is readily feasible and can provide a valuable guide to the choice of crystals suitable for X-ray structure analysis. Such evaluations seem to be a prerequisite for the design of mechanisms based on enzyme structure.

CHEMICAL MODIFICATIONS TO PROBE THE ACTIVE SITES OF ENZYMES

Monoarsanilazotyrosyl-248 carboxypeptidase

Comparison of the kinetic characteristics of enzyme crystals with those in solution constitutes only one approach to the study of the relationship of conformation to function. We have also searched for other means that could *simultaneously* gauge activity and the structural dynamics of the active centre in solution in a wide range of environmental conditions. Several chromophoric probes have been found helpful in this regard, particularly azotyrosyl and nitrotyrosyl residues. The location of Tyr-248 relative to the active-site zinc atom has been thought to be critical for relating the three-dimensional structure of carboxypeptidase A to its catalytic mechanism (Lipscomb et al. 1968; Quiocho et al. 1972). Coupling of Tyr-248 with diazotized arsanilic acid generates a visible absorption spectrum with a maximum at 510 nm that has been shown to be due to an intramolecular coordination complex between the monoarsanilazotyrosyl residue and zinc (Johansen & Vallee 1971, 1973). This complex is disrupted by crystallization of the enzyme, by removal of the metal, or by addition of substrates or inhibitors (Johansen & Vallee 1975). Absorption and circular dichroism–pH titrations of the modified zinc enzyme demonstrate two pK_{app} values, 7.7 and 9.5, characterizing the formation and dissociation of the complex, respectively. Titrations of the apoenzyme, which completely lacks the 510 nm absorption band at all pH values, reveal a single pK_{app}, 9.4, due to the ionization of the azophenol (λ_{max} 485 nm). Substitution of other metals for zinc results in analogous complexes with absorption maxima and circular dichroism extrema characteristic of the particular metal. Studies with nitrocarboxypeptidase (Riordan & Muszyńska 1974) also provide evidence consistent with the interaction of tyrosine-248 and the zinc atom.

Resonance Raman spectroscopic studies of arsanilazotyrosyl-248 carboxypeptidase: structure of the intramolecular coordination complex

We have recently examined the structural details of the azoTyr-248–Zn complex by means of resonance Raman (rR) spectroscopy (Scheule *et al.* 1977). The rR spectrum of arsanilazotyrosyl-248 carboxypeptidase A exhibits only the vibrational bands of its chromophoric azoTyr-248 residue, unhindered by background interference from either water or other components of the protein. The spectrum contains multiple, discrete bands which change as a function of pH, consistent with the existence of interconvertible species of the azotyrosine probe in solution (Fig. 1). Spectra of model azophenols and of the apoazoenzyme establish the identity of these species. All conclusions about the azoenzyme based on the rR spectra, including the apparent pK values for the interconversions, agree with those drawn earlier from studies by absorption spectroscopy. In addition, the properties of rR bands that have been assigned to the motions of the specific atoms of the complex on the basis of isotope substitutions provide details of the interactions of those atoms with the active zinc atom.

$\rightarrow$

Fig. 1. (A) Absorption spectra of arsanilazotyrosyl-248 carboxypeptidase A at the pH values indicated. (B) Resonance Raman spectra of the three species of arsanilazotyrosyl-248 carboxypeptidase A at the pH values indicated. At intermediate pH values, the spectra constitute a superposition of these three basic spectra (from Scheule *et al.* 1977).

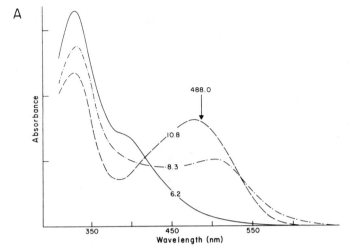

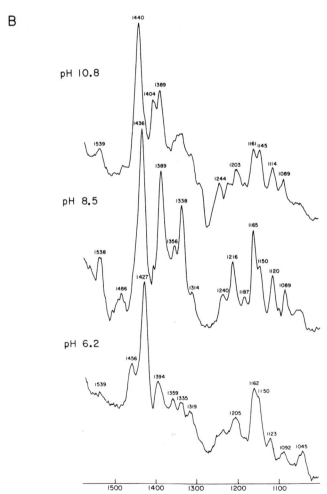

Fig. 1.

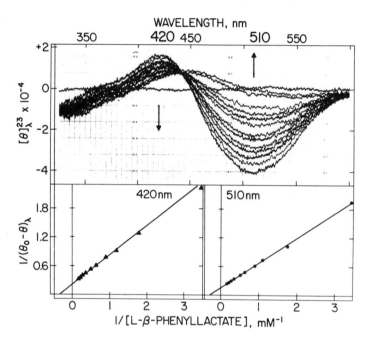

FIG. 2. Top: Effect of L-phenyllactate on the circular dichroic spectrum of zinc azoTyr-248 carboxypeptidase in 0.05M-Tris–0.5M-NaCl, pH 8.5, 23 °C, uncorrected for dilution. L-Phenyllactate concentration is varied from 0 to 30 mmol/l. Arrows indicate the direction of change at the corresponding wavelength set in larger ciphers. Bottom: Double reciprocal plots of $1/(\theta_0 - \theta)_\lambda$ against $1/$[L-phenyllactate] at (right) 510 nm ($\bullet$)   (K_{app} 2.8 mmol/l) and (left) 420 nm ($\blacktriangle$) (K_{app} 2.5 mmol/l), both calculated from the circular dichroic spectral titrations in the upper panel (from Johansen *et al.* 1976).

Detection of multiple binding modes at the active site of carboxypeptidase A by circular dichroism–inhibitor titrations

The azoTyr-248 group constitutes two discrete probes. One signals perturbations of the azoTyr-248 itself and the other of the azoTyr-248–Zn chelate. Each can detect the same and/or different interactions and environmental changes, greatly magnifying the scope of this optically active chromophore to reveal important aspects of the topography of the active centre and its interaction with inhibitors and substrates in solution.

The response of the circular dichroic spectrum of the azoenzyme to titration with a series of agents known to inhibit native carboxypeptidase in different modes has been examined in some detail (Johansen *et al.* 1976). Titrations of the azoenzyme with a series of competitive inhibitors (e.g., L-benzylsuccinate, L-phenylalanine and L-phenyllactate; Fig. 2), and the pseudosubstrate, Gly-L-Tyr, generate characteristic circular dichroic spectra. Analyses of these

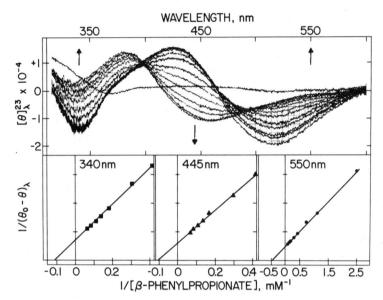

FIG. 3. Top: Effect of β-phenylpropionate on the circular dichroism spectrum of zinc azoTyr-248 carboxypeptidase in 0.05M-Tris–0.5M-NaCl, pH 7.6, 23 °C. β-Phenylpropionate concentration was varied from 0 to 91 mmol/l. The spectra are not corrected for dilution, which is 18% for the final spectrum. Arrows indicate the direction of change at the corresponding wavelength, set in larger ciphers. Bottom: Double reciprocal plots of $1/(\theta_0 - \theta)_\lambda$ against $1/[\beta$-phenylpropionate] at (right) 550 nm (●) (K_{app} 2.4 mmol/l), (middle) 445 nm (▲) (K_{app} 9.3 mmol/l) and (left) 340 nm (■) (K_{app} 9.4 mmol/l), calculated from the circular dichroic spectral titrations in the upper panel (from Johansen *et al.* 1976).

titrations reveal a single binding constant for each agent indicating that only one molecule binds to the active centre. Other agents, e.g., β-phenylpropionate (Fig. 3) and phenylacetate, whose mixed modes of inhibitition have previously been resolved into competitive and non-competitive components (Auld *et al.* 1972), cause different spectral effects. Their circular dichroic–titration curves are consistent with two molecules of inhibitor binding to the enzyme, as inferred from both equilibrium and kinetic studies. The interactions leading to competitive and non-competitive inhibition, respectively, can be differentiated by their characteristic effects on the extrema at 340 and 420 nm, reflecting azoTyr-248, and the negative 510 nm circular dichroism band, due to the chelate with zinc. Non-competitive inhibitors and modifiers induce yet additional spectral features, each signalling changes in a particular active centre environment. The general as well as the specific features of these circular dichroic titrations are consistent with our mechanistic views (Vallee *et al.* 1968).

Monitoring catalytic events

It is apparent from the above studies that the azoTyr-248–Zn complex is an extremely effective probe for exploring the relationship between catalytic activity and local structure by equilibrium approaches. The investigation of local changes in conformation that are essential to and synchronous with catalysis requires a probe with similarly specific characteristics and extraordinarily fast but kinetically-detectable response times. We have been able to use the same site-specific enzyme modifications as sensors of structural changes accompanying catalysis in kinetic studies. The spectrum of the intense, chromophoric complex between azoTyr-248 and the zinc atom responds dynamically to environmental factors as well as the physical state of the enzyme.

We have examined the kinetics of association and dissociation of the azo-Tyr-248–Zn complex by stopped-flow pH and temperature-jump methods (Harrison *et al.* 1975). The rate constant for the dissociation process, 64 000 s^{-1}, is orders of magnitude greater than that for the catalytic step itself (about 0.01–100 s^{-1}). Rapidly-turned-over peptide and ester substrates disrupt the azoTyr-248–Zn complex before hydrolysis occurs. As a consequence, formation of uncomplexed azoTyr-248, binding of substrate and catalysis can all be monitored independently. The results of these dynamic studies specify a course of catalytic events, different from that postulated based on X-ray structure analysis. Tyr-248 is displaced, but away from the zinc and, hence, in the direction opposite to the inward movement postulated on the basis of X-ray studies (Lipscomb *et al.* 1968).

AzoTyr-248 carboxypeptidase has all the features which are essential for mechanistic studies: it is enzymically active; the spectra of the intramolecular metal complex differ characteristically from those of its constituents; it responds dynamically to environmental factors; and the response time of the probe itself is rapid enough to measure the catalytic step. These combined kinetic and spectral properties of the metal complex render it a potent spectro-kinetic probe for observing microscopic details of the catalytic process.

Monitoring multiple conformational states in solution

The azoTyr-248–Zn complex also provides a means to detect multiple conformations of the enzyme in solution by stopped-flow pH and temperature-jump experiments (Harrison *et al.* 1975). On mixing azoTyr-248 carboxypeptidase at pH 8.5 with pH 6.0 buffer, 96% of the red azoTyr-248–Zn complex is converted into the yellow azophenol within 3 ms, the mixing time

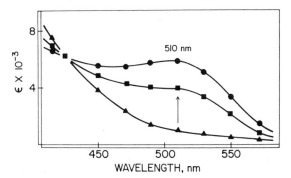

FIG. 4. Spectra of azoTyr-248-carboxypeptidase A, reconstructed from stopped-flow pH jump experiments from pH 6.5 to 8.2 at 20 nm intervals. The spectrum at zero time (▲) changes in 3 ms to the spectrum of 60% of the azoTyr-248–zinc complex (■) followed by a slower exponential change (k = 2.2 s^{-1}) finally to reach the equilibrium spectrum (●) of azoTyr-248–Zn complex. Final conditions: enzyme, 25 μmol/l, pH 8.2, 50μM-Hepes, 1M-NaCl, 25 °C (from Harrison *et al.* 1975).

of the stopped-flow instrument. The rate constant for the rapid process, determined by temperature-jump analysis, is pH-dependent and varies from 100 000 to 50 000 s^{-1}. The remaining 3% of the total absorbance change occurs slowly, with a first-order rate constant of 0.4 s^{-1}. This slow change implies a relaxation process involving two different conformations of yellow azocarboxypeptidase.

Jumping from pH 6 to 8.2 generates only about 60% of the final absorbance at 510 nm within the 3 ms mixing time of the experiment. The remaining 40% of the colour appears with a first-order rate constant of 5.0 s^{-1}. The reconstruction of the azoTyr-248-carboxypeptidase spectra in pH jump experiments from pH 6.5 to 8.2 identifies a time-dependent conformational step (Fig. 4). This again suggests the presence of two yellow species, one that forms the metal complex readily on raising the pH and another that does not. Varying the initial pH of the jump experiments while keeping the final pH constant demonstrates that the interconversion of the two yellow species is independent of pH.

The mechanism governing these interrelationships involves two distinct processes. The first, $E_r + H^+ \leftrightharpoons E_y$ (with an equilibrium constant, K_1) is extremely rapid (k_{fast} about 10^5 s^{-1}) and reflects the pH-dependent dissociation of the metal complex. The second, $E_y \leftrightharpoons E_{y'}$, is much slower ($k_{slow}$ = 5.0 s^{-1}) and is due to the pH-independent interconversion of two distinct populations of protein molecules, E_y and $E_{y'}$, in which the yellow azoTyr-248 is in different conformations. These two separate processes can be recognized readily since the first is three to four orders of magnitude faster than the second. Even though both E_y and $E_{y'}$ are yellow and, hence, cannot be directly differentiated by their spectra, the amplitudes of the stopped-flow pH-jump measurements are directly related to their concentrations. Therefore, the relaxation process becomes a precise gauge of their interconversion. Jumping from pH 6.0—at which only the two yellow forms of the enzyme are present and in equilibrium— to pH 8.5 reveals that about 60% of the molecules are in the E_y form and can complex zinc rapidly. Conversely, on jumping to pH 6.0 from pH 8.5, almost all the colour change occurs rapidly, reflecting the disruption of the red complex and the equilibration of E_r and E_y as governed by K_1. The remaining colour change occurs as E_y slowly equilibrates with $E_{y'}$ thus allowing further conversion of E_r into E_y. The best fit of all experimental data to theoretical curves is obtained with a pK_1 of 7.7, identical with the value derived from spectral equilibrium data (Johansen & Vallee 1973, 1975).

In a previous stopped-flow pH-jump experiment only a single process with a rate constant of about 6 s^{-1} was recognized (Quiocho *et al.* 1972). This study failed to detect the much more rapid process and, as a consequence, led to several postulates to account for those observations. Obviously, these earlier conjectures are no longer pertinent.

These studies point to the existence of rapidly-equilibrating substructures of carboxypeptidase A, consistent with the view that in solution enzymes can adopt multiple, interconvertible, related conformations that could either facilitate or impede catalysis. In crystals, rearrangements of molecular structure could be severely impaired or restricted, and crystallization might single out either active or inactive conformations. It is not surprising that in some cases crystals could have greatly reduced activities or otherwise markedly altered catalytic behaviour, as observed for carboxypeptidase A.

CHROMOPHORIC METAL ATOMS AS ACTIVE-SITE PROBES

The presence of a chromophoric metal atom at the active site of a metallo-enzyme provides another approach to the detection and study of local conformational changes that might occur during the course of catalysis. The metal atom of zinc metalloenzymes is diamagnetic and does not absorb radiation in the visible spectral region. However, metals that are chromophoric, particularly cobalt, can be substituted for zinc to give catalytically-active derivatives (Vallee & Latt 1970; Lindskog & Nyman 1964; Simpson & Vallee 1968; Sytkowski & Vallee 1976). Detailed analyses of the absorption, circular dichroic, magnetocircular dichroic and e.p.r. spectra of cobalt metalloenzymes define features of metal coordination. Such spectra

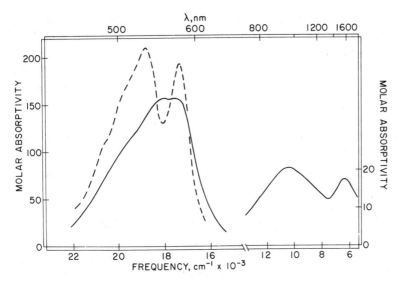

FIG. 5. Absorbtion spectra of cobalt carboxypeptidase. For the visible region the enzyme (1mmol/l) was dissolved in 1M-NaCl, 0.05M-Tris–HCl, pH 7.1, 20 °C (—). Another enzyme solution (about 3 mmol/l) was diluted with glycerol to 45% (v/v) and cooled to 4.2 K (---) for spectral measurements. For the near infrared region apocarboxypeptidase (1.5 mmol/l) was dissolved in 1M-NaCl, 0.005M-[D]Tris–Cl in D_2O, pD 7.2. The sample cuvette contained 1.5mM-enzyme plus $CoSO_4$ in D_2O buffer to give a final total cobalt concentration of 2.0 mmol/l and hence a 0.5mM-excess of free Co(II) ions: the reference cell contained 1.5mM-apoenzyme brought to volume with buffer.

are sensitive to environmental changes induced by inhibitor or substrate binding and can therefore signal the formation of intermediates during catalysis.

Thus, for example, cobalt carboxypeptidase A is fully active both as a peptidase and an esterase. Its visible absorption spectrum (Fig. 5) has a shoulder near 500 nm and maxima at 555 and 572 nm, both with absorptivities of about 150 and its infrared spectrum has bands at 940 and 1510 cm^{-1} (ε ~20). Increased resolution of the visible bands but no reduction in absorptivity occurs at 4 K. Overall the spectrum is indicative of an irregular coordination geometry and tight bonding (Latt & Vallee 1971). It is entirely dependent on the three-dimensional structure of the protein, since it is abolished by the addition of denaturing agents. It has been suggested that the metal and its ligands might constitute part of a system poised to act in catalysis (Vallee & Williams 1968). These data are consistent with such a hypothesis.

The circular dichroic spectrum of the enzyme has a negative ellipticity band at 538 nm and a shoulder at about 500 nm. Only the absorption band at 572 nm becomes optically active in a magnetic field. Many inhibitors markedly perturb the circular dichroic spectrum as do pseudosubstrates such as glycyl-

L-tyrosine. These striking changes suggest that significant electronic re-arrangements occur about the cobalt atom on interaction with substrates which reflect the role of the metal in peptide hydrolysis. The metal is thought to polarize the carbonyl group of the peptide bond that is destined for cleavage. It coordinates directly with the carbonyl oxygen atom to form an inner-sphere complex (Vallee *et al.* 1963; Lipscomb *et al.* 1968). The spectral data suggest that this interaction profoundly influences the metal environment. However, the absorption and magnetocircular dichroic spectra indicate that the basic metal coordination is unchanged on formation of the complex.

Evidence for the inner-sphere nature of the enzyme–substrate complex comes from oxidation studies in which the substitution-labile Co(II) is converted into substitution-inert Co(III). Initial attempts to do such experiments with H_2O_2 as the oxidizing agent (Kang *et al.* 1975) were later shown to be invalid. In other instances, such as alkaline phosphatase of *E. coli*, H_2O_2 has proven to be satisfactory for this purpose, but with carboxypeptidase it causes loss of cobalt and induces other artifacts that obviate evaluation. However, Van Wart & Vallee (1977) found that *m*-chloroperbenzoic acid, a competitive inhibitor analogue, can selectively convert the Co(II) into the Co(III) enzyme with concomitant abolition of both activities. Evidence for the formation of a Co(III) enzyme was obtained by absorption and e.p.r. spectroscopy, by gel filtration, and by studying rates of metal dissociation. Loss of esterase activity was shown to be due to loss of substrate binding, whereas peptides could still bind to the Co(III) enzyme but were not hydrolysed. Importantly, peptide substrates do not undergo a single turnover, thus ruling out the possibility of an outer-sphere hydrolytic mechanism or a so-called metal-hydroxide mechanism (Lipscomb *et al.* 1968; Wells & Bruice 1977).

Metal → metal resonance energy transfer

Cobalt substitution can also be used to probe topography beyond the immediate environment of the active site. Thermolysin from *B. thermopro-teolyticus* contains four calcium atoms in addition to the single zinc atom at the active site. Replacement of zinc by cobalt yields a product with twice the native activity (Holmquist & Vallee 1974). In addition, three of its four calcium sites can be occupied by trivalent lanthanide ions thereby excluding all calcium. X-ray structure analysis shows that the most readily substituted calcium site is only 1.37 nm from the active-site zinc- (or cobalt-) binding site. Certain lanthanides, for example Tb(III) and Eu(III), exhibit markedly enhanced fluorescence when bound to proteins, and Tb(III) fluorescence overlaps the absorption of the cobalt chromophore. Hence Co(II)- and Tb(III)-

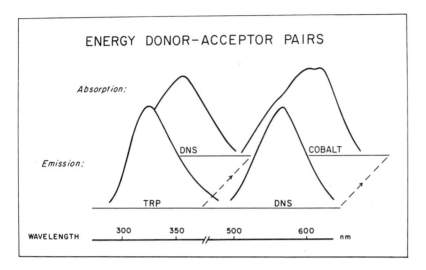

FIG. 6. Schematic representation of the overlap relationships between enzyme tryptophan, substrate dansyl, and cobalt absorption and emission bands that constitute the energy donor–acceptor relay system critical to observation of the E–S complexes.

substituted thermolysin favour inter-ion resonance-energy transfer and can be used for distance measurements by this technique (Horrocks *et al.* 1975). Thus, addition of Tb(III) to zinc or zinc-free thermolysin can be monitored by exciting at 280 nm and observing the Tb(III) fluorescence emitted at 545 nm. Addition of Co(II) to the zinc-free enzyme decreases the terbium fluorescence intensity by 89.5%. Treatment with *N*-bromosuccinimide quenches enzymic tryptophyl and Tb(III) fluorescence to a similar extent and suggests the operation of a tryptophan → Tb(III) → Co(II) resonance-energy relay system in the enzyme. This type of radiationless energy transfer between two different metal atom sites of a protein provides a new class of probes to measure intramolecular distances of enzymes in solution and can be used to monitor conformational changes that occur during catalysis. The distance between the Tb(III) and Co(II) atoms measured by this means coincides exactly with that for the corresponding Ca(II) and Zn(II) sites of the native enzyme, as gauged from X-ray structure analysis (Horrocks *et al.* 1975).

Syncatalytic measurement of distances

The fluorescent transfer of electronic excitation energy can be detected rapidly enough to allow examination of syncatalytic events. We have taken advantage of this by using the spectral overlap of dansyl fluorescence and cobalt absorption. The detectability of the dansyl group is well known in peptide chemistry and has been a mainstay in sequence analysis for many years. It can also serve as an NH_2-terminal blocking group for peptide substrates of carboxypeptidase and other endo- or exo-peptidases. We have used such N-dansylated oligopeptides and ester substrates to measure distances between the cobalt atom of cobalt carboxypeptidase and the substrate dansyl group within the enzyme active centre while simultaneously signalling other aspects of active-centre topography as the enzymic reaction is in progress (Latt *et al.* 1970). The relay system, analogous to the metal–metal energy transfer just described, involves enzyme tryptophyl residues, the dansyl group of bound–but not of free–substrate, and the cobalt atom. The overlap of the fluorescence emission and absorption spectra is shown in Fig. 6. In this system the dansyl group plays a dual role: subsequent to excitation of tryptophan, energy is transferred in tandem to the dansyl group and from the dansyl group to the cobalt atom. Dissociation of the product after scission of the susceptible bond terminates this energy relay system (Latt *et al.* 1970, 1972). It should be noted that energy transfer between tryptophyl residues of the enzyme and the dansyl group of bound substrate allows observations of the formation and breakdown of the E–S complexes for both the zinc and cobalt enzymes, irrespective of the degree of energy transfer between the dansyl group and the cobalt atom (*vide infra*).

Binding of dansyl substrates to the cobalt enzyme is accompanied by quenching of enzyme tryptophyl fluorescence. The bound-dansyl fluorescence is quenched also owing to energy transfer to the cobalt atom. The degree of quenching of the dansyl emission is a sensitive function of its distance from the cobalt atom. When the dansyl group is relatively close to the cobalt atom, as for the dipeptide Dns-Gly-Phe, its emission is quenched almost totally. Insertion of additional amino acid residues to give tri- and tetra-peptides, e.g. Dns-[Gly]$_2$-Phe or Dns-[Gly]$_3$-Phe, moves the dansyl group progressively farther away from the cobalt atom, and the degree of dansyl quenching decreases in proportion to the inverse sixth power of the distance between them, in accord with Förster's formulation of resonance energy transfer.

Energy transfer, T, from the dansyl group of the substrate in the E–S complex to the cobalt atom of the enzyme is calculated from the relative fluorescence efficiency, F'_{Co}/F'_{Zn}, of the dansyl group of the substrate when bound to either of the two enzymes (equation 1). The ratio of donor-acceptor

$$1 - T = F'_{Co}/F'_{Zn} \tag{1}$$

separation, R, to the critical distance for 50% efficient energy transfer, R_o, calculated from the Förster equation (1948, 1965) is given by equations (2)

$$1 - T = 1/[1 + (R_0/R)^6] \tag{2}$$

$$R_0^6 = 8.78 \times 10^{-25} (x^2 Q J / n^4) \tag{3}$$

and (3) where the donor quantum yield in the absence of transfer, Q, and the overlap integral, J, are quantities determined experimentally; the index of refraction of the solvent, n, is 1.33; and a value of 2/3 is employed for the random dipole orientation factor x^2 (Latt *et al.* 1970).

Assignment of the probable value of x^2 has consistently posed problems in the determination of R: x^2 can range from 0, when all vectors are mutually orthogonal, to 4, when all vectors are parallel. The random average for donor-acceptor orientation is 2/3. The choice of cobalt as an energy acceptor has the great advantage that the nearly triply-degenerate visible-absorption transitions of cobalt limit the possible range of x^2 values from 1/3 to 4/3; 2/3 still remains the average, but the probability that this is valid is greatly enhanced for this case (Latt *et al.* 1970, 1972).

The distances between the cobalt atom, acting at the susceptible peptide bond, and the dansyl group of bound substrates, as determined experimentally by means of energy transfer (Latt *et al.* 1970, 1972), are within the limits of those measured on Corey–Pauling–Koltun models of such peptides, assumed to be in an extended conformation (Table 1). Further, the increases in distance as a function of the chain length of the peptides are internally consistent for both the Phe and Trp sets of substrates. Moreover, the distances measured for corresponding members of the two sets of peptides agree well.

TABLE 1

Dansyl–cobalt distances in carboxypeptidase–substrate complexes[a]

Substrate	R/nm[b]	R/nm from C–P–K models[c]
Dns-Gly-L-Phe	>0.8	0.7
Dns-Gly-L-Trp	>0.8	
Dns-Gly-Gly-L-Phe	1.11–1.13	1.0
Dns-Gly-Gly-L-Trp	1.08–1.23	
Dns-Gly-Gly-Gly-L-Phe	1.17–1.27	1.3
Dns-Gly-Gly-Gly-L-Trp	1.29–1.44	
Dns-Gly-Gly-Gly-Gly-L-Phe	1.41–1.47	1.6

[a] 1M-NaCl–0.02M-Tris, pH 7.5, 25 °C (Latt *et al.* 1972).
[b] The range of experimental values for the distance, R, reflects that of possible bound-substrate quantum yields.
[c] Measured on Corey–Pauling–Koltun molecular models from the centre of the dansyl group to the cobalt atom, assuming the peptide to be in an extended conformation and the metal to bind the oxygen atom of the COOH-terminal peptide bond.

These dansyl–cobalt energy transfer measurements determine the radii of arcs about the cobalt atom along which the dansyl group might lie. The intersection of these arcs with the enzyme surface define the regions surveyed by the dansyl peptide substrates, but substrate orientation within such contours remains to be determined.

It is possible also to determine the average distance of the dansyl group from responsive enzyme tryptophyl residues; such measurements would be most meaningful in enzymes containing only one tryptophyl residue. Modification of any of the components of the energy relay system, either in the enzyme, in the substrate or in both can extend the approach. Numerous potential combinations of placing donor and acceptor groups at various different strategic positions on the enzyme surface are feasible to measure catalysis-related distances and perhaps even the velocity of conformational movement. Even though the present approach already allows qualitative decisions concerning the polarity of the active centre environment of this and various other enzyme systems, the scope of the information can be increased further by complementing fluorescence intensity with polarization and lifetime measurements.

Detection of intermediates

As stated above, even in the zinc enzyme resonance-energy transfer between fluorescent enzyme tryptophyl residues serving as intrinsic donors and an extrinsically placed dansyl group acceptor in the substrate allows direct observation of the enzyme–substrate complex. The spectral overlap between the dansyl group absorption and tryptophan emission is excellent. Quantitatively, the degree of energy transfer is sensitive to the distance between and orientation of the donor-acceptor pair and to the environment of the acceptor. Differences in tryptophan to dansyl transfer efficiencies and/or dansyl quantum yields can characterize the resultant E–S species. Thus, if a set of reversible enzyme–substrate complexes, E–S, and/or a covalent intermediate, EA, are formed in the course of an enzyme-catalysed reaction, as in the following scheme:

$$E + S \rightleftharpoons (ES)_1 \rightleftharpoons (ES)_2 \rightleftharpoons (ES)_n \rightleftharpoons EA + P_1 \rightleftharpoons E + P,$$

it is possible to determine the minimal number of significantly populated states, the rates of interconversion of molecules among them, and the equilibrium constants determining the relative proportion of the populations.

The rapid equilibrium of carboxypeptidase A and Dns-[Gly]$_3$-L-OPhe to

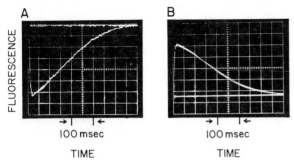

FIG. 7. Enzyme tryptophan (A) and substrate dansyl (B) fluorescence during the time course of zinc carboxypeptidase (2.5 μmol/l) catalysed hydrolysis of 0.1mM-Dns-[Gly]$_3$-L-OPhe in NaCl–0.03M-Tris, pH 7.5, 25 °C. The fluorescence of either tryptophan (A) or dansyl (B) was measured as a function of time under stopped-flow conditions. Oscilloscope traces of duplicate reactions are shown in each case. Excitation was at 285 nm. Enzyme tryptophan fluorescence was measured by means of band-pass filter peaking at 360 nm and dansyl emission was measured by a 430 nm cut-off filter. Scale sensitivities for (A) and (B) are 50 mV/div and 500 mV/div. The existence of the E–S complex is signalled either by (A) the suppression of enzyme tryptophan fluorescence (quenching by the dansyl group) or (B) enhancement of the substrate dansyl group fluorescence (energy transfer from enzyme tryptophan).

form the E–S complex (Fig. 7) and previous kinetic data for blocked oligopeptide substrates indicate that carboxypeptidase acts entirely in accord with the classical Michaelis–Menten kinetic scheme:

$$E + S \underset{}{\overset{K_s}{\rightleftharpoons}} E\text{--}S \overset{k_2}{\rightarrow} P + E$$

It should be noted, however, that at 25 °C, the temperature used for these studies, it is possible for the initial signal to reflect a distribution of E–S complexes that have reached equilibrium within the mixing time of the instrument. Lowering the temperature in combination with rapid mixing techniques should allow the detection of such rapidly equilibrating species.

This approach has also been applied successfully to delineate mechanisms of inhibition of carboxypeptidase A (Auld *et al.* 1972), the enzymic consequences of metal substitution and differentiation of the mechanism of ester and peptide hydrolysis (Auld & Holmquist 1974). More recently it has been applied to several other hydrolytic enzymes, such as yeast carboxypeptidase, chymotrypsin and alkaline phosphatase. The intermediates observed have allowed quantitative assessment of kinetic schemes for these enzymes.

Kinetic studies of carboxypeptidase A over the past two decades have revealed complexities in its catalytic behaviour toward acylamino acids, dipeptides and their ester analogues. The varying degrees of activation and inhibition imposed by these substrates and their products of hydrolysis have

TABLE 2

Carboxylic acid inhibitors of the carboxypeptidase A-catalysed hydrolysis of esters and peptides[a]

Substrate	Inhibitor	$10^{-4}K_I/(mol/l)$	Type of inhibition
Dns-[Gly]$_3$-L-Phe	Phenylacetate	3.3	Non-competitive
Dns-[Gly]$_3$-L-OPhe	Phenylacetate	3.2	Competitive
Bz-[Gly]$_2$-L-OPhe	β-Phenylpropionate	1.2	Non-competitive
Bz-[Gly]$_2$-L-Phe	β-Phenylpropionate	1.2	Competitive
Cbz-[Gly]$_2$-L-Phe	Indole-3-acetate	1.7	Non-competitive
Bz-[Gly]$_2$-L-OLeu	Indole-3-acetate	1.6	Competitive

[a] Assays performed at 25 °C, pH 7.5, 1.0M-NaCl, 0.05M-Tris, except for the phenylacetate study where the conditions were pH 6.5, 1.0M-NaCl, 0.03M-Mes buffer (Auld & Holmquist 1974).

precluded the determination of meaningful kinetic constants. A schematic model, based on multiple modes of substrate binding, has been proposed to accommodate the kinetic characteristics of carboxypeptidase towards these substrates (Vallee *et al.* 1968). It incorporates non-identical but overlapping binding of dipeptides and analogous esters as the basis for the kinetics anomalies. The model suggested that increasing the length of these substrates would eliminate anomalous binding and likely normalize their enzymic hydrolysis. Our kinetic studies of both oligopeptides and their ester analogues have, indeed, borne out this prediction (Auld & Vallee 1970). However, the question has remained whether or not the native and modified enzymes which discriminate between peptide and ester substrates displaying anomalous kinetics (e.g., Cbz-Gly-L-Phe and Bz-Gly-L-OPhe) would also differentiate between members of peptide-ester pairs obeying Michaelis–Menten kinetics. Studies of oligopeptides and their depsipeptide analogues answer this question. Auld & Holmquist (1974) have obtained several types of evidence indicating unequivocally that the overall mechanisms of the carboxypeptidase A-catalysed ester and peptide hydrolysis must differ. First, carboxylic acids such as β-phenylpropionate and phenylacetate inhibit hydrolysis of peptides non-competitively but that of their exact ester analogues competitively (Table 2). The identity of the K_I values obtained for both classes of substrates suggest that both ester and peptide substrates form identical E–I complexes, but the different modes of inhibition show that they must bind differently.

One of the earliest pieces of information demonstrating a difference in the esterase and peptidase activities of the enzyme was obtained by substitution of Cd for Zn. The resulting cadmium carboxypeptidase retains activity towards Bz-Gly-L-OPhe but is almost inactive toward Cbz-Gly-L-Phe. However, the peptide Cbz-Gly-L-Phe inhibits this esterase activity, i.e. the cadmium enzyme can still bind the peptide.

Stopped-flow fluorescence studies of E–S complexes provide a direct

TABLE 3

Hydrolysis of matched ester and peptide pairs by zinc carboxypeptidase [(CPD)Zn] and cadmium carboxypeptidase [(CPD)Cd][a]

Substrate	[(CPD)Zn]		[(CPD)Cd]	
	k_{cat}/min^{-1}	$10^4 K_m/(mol/l)$	k_{cat}/min^{-1}	$10^4 K_m/(mol/l)$
Bz-[Gly]$_2$-L-OPhe	30 000	3.3	34 000	79.0
Bz-[Gly]$_3$-L-OPhe	31 000	3.4	45 000	29.00
Dns-[Gly]$_3$-L-OPhe	11 000	0.25	28 000	2.2
Bz-[Gly]$_2$-L-Phe	1 200	10.0	41	8.0
Bz-[Gly]$_3$-L-Phe	2 600	37.0	86	41.0
Dns-[Gly]$_3$-L-Phe	4 200	8.0	400	8.1

[a] Assays performed at 25 °C, pH 7.5, 1.0M-NaCl, and a buffer concentration of 0.05M-Tris for peptide hydrolysis and 10^{-4}M-Tris for ester hydrolysis. The carboxypeptidase A (Anson) was used in all cases.

TABLE 4

Metallocarboxypeptidase-catalysed hydrolysis of Bz-[Gly]$_2$-L-Phe and Bz-[Gly]$_2$-L-OPhe[a]

Metal	k_{cat}/min^{-1}	$10^{-3} K_m/(l/mol)$	$10^{-4} k_{cat}/min^{-1}$	$K_m^{-1}/(l/mol)$
Cobalt	6000[b]	1.5[b]	3.9	3300
Zinc	1200	1.0	3.0	3000
Manganese	230[b]	2.8[b]	3.6	660
Cadmium	41	1.3	3.4	120

[a] Assays performed at 25 °C, pH 7.5, 1.0M-NaCl, and a buffer concentration of 0.05M-Tris for peptide hydrolysis and 10^{-4}M-Tris for ester hydrolysis.
[b] Values are for carboxypeptidase A (Cox) (Auld & Vallee 1970). All other values are for carboxypeptidase A (Anson).

comparison of the peptide-binding affinities of the zinc and cadmium enzymes and, simultaneously, an explanation for the different roles of metals in peptide and ester hydrolysis. Cadmium carboxypeptidase binds the peptide Dns-[Gly]$_3$-L-Phe as readily as does zinc carboxypeptidase but catalyses its hydrolysis at a considerably reduced rate, in agreement with steady-state rate studies of oligopeptides. The catalytic rate constants of the Cd enzyme are markedly decreased for all peptides examined, but the association constants (K_m^{-1} values) of the Cd enzyme are identical to those of the Zn enzyme. In marked contrast, the catalytic rate constants of the Cd enzyme for esters are nearly the same as those of the Zn enzyme but the association constants of the Cd enzyme are greatly decreased (Table 3).

Substitution of cobalt or manganese for zinc also affects peptide and ester hydrolysis (Table 4). The k_{cat} values for Bz-[Gly]$_2$-L-Phe hydrolysis follow the order Co > Zn > Mn > Cd, cobalt carboxypeptidase being 150 times more active than cadmium carboxypeptidase. However, the corresponding association constants, K_m^{-1}, are essentially identical. On the other hand, metal substitution markedly alters the binding affinity of the exact ester analogue, Bz-[Gly]$_2$-L-OPhe. The K_m^{-1} values decrease in the same order as do the k_{cat} values of peptide hydrolysis, i.e., Co > Zn > Mn > Cd, the affinity of the ester for cobalt carboxypeptidase being 30 times greater than for cadmium carboxypeptidase. Metal substitution, however, has no significant effect on the catalytic rate constant for ester hydrolysis.

Stopped-flow fluorescence studies show that the apoenzyme binds peptides as tightly as does the zinc enzyme; their initial E–S complex concentrations are nearly equal. However, the apoenzyme–peptide complex is stable and does not break down to products during the time needed for complete hydrolysis of the peptide by the zinc enzyme. Hence, removal of the metal does not alter the affinity of the enzyme for the peptide. Thus, although the apoenzyme cannot catalyse the hydrolysis of peptide substrates, it binds them to the same degree as does the zinc enzyme. However, removal of zinc from the native enzyme decreases binding of the exact ester analogue by an order of magnitude. Similarly, as discussed above, the cobalt(III) enzyme, which is inactive toward both peptide and ester substrates (Van Wart & Vallee 1977), binds peptides but not esters. All these studies indicate that the binding of peptides to metallocarboxypeptidases must differ from that of esters.

A positively-charged residue in the active centre has long been thought to be a major determinant of the specificity of carboxypeptidase for which a free COOH-terminal carboxy group of the substrate is mandatory (Wald-schmidt-Leitz 1931; Smith 1949; Vallee et al. 1963). Since acylation experiments have excluded lysines (Riordan & Vallee 1963), it seemed possible that an arginyl residue might function as the binding locus for the carboxylate group (Vallee & Riordan 1968). Based on X-ray crystallographic studies of the Gly-L-Tyr complex of the crystalline enzyme it has since been concluded that the carboxy group of this pseudosubstrate is bound to Arg-145. It has been inferred, further, that this arginine is the determinant of enzyme specificity for interaction with the COOH-terminal free carboxy group of *all* substrates and is thus indispensable to catalysis by carboxypeptidase (Lipscomb et al. 1968). Chemical modification of arginine in carboxypeptidase diminishes the rate of Cbz-Gly-L-Phe hydrolysis, consistent with the hypothesis that peptide substrates can bind to an arginyl residue (Riordan 1973). However, this modification does not diminish esterase activity toward Bz-Gly-L-OPhe suggesting that a positively-charged group other than arginine serves as the recognition site and binding locus for the free carboxy group of esters.

Direct binding of substrates to the metal through their carboxy group was first suggested by Lumry & Smith (1955). In view of the above results, the active-site metal may serve as a primary binding locus for esters, but such binding appears unlikely for peptides. Direct binding studies and metal-ion-exchange data led to the proposal (Coleman & Vallee 1962, 1964) that binding of β-phenylpropionate involves an interaction of the carboxy group with the metal. Since then crystallographic studies with p-iodo-β-phenylpropionate (Navon et al. 1970) have confirmed this postulate. Presumably, the carboxy groups both of esters and of such carboxylic acid inhibitors compete for the metal whereas peptides seem to bind to the guanido group of an arginyl residue (Riordan 1973).

X-ray diffraction studies of the binding of ester substrates or ester pseudo-substrates to the crystalline enzyme have not been successful. However, a coordinated series of conformational changes is thought to occur when the peptide pseudosubstrate, Gly-L-Tyr, binds to the crystalline enzyme (Lipscomb et al. 1968). The X-ray structure analysis of the non-productive enzyme–Gly-Tyr complex led to the suggestion that an initial interaction of the carboxy group of Gly-Tyr with Arg-145 brings about a substrate-induced conformational change of 1.2 nm in Tyr-248 so that the phenolic hydroxy group would move into the vicinity of the susceptible bond of the substrate (Lipscomb et al. 1968). Two facts are at variance with this interpretation. First, as indicated above, Tyr-248 can be coordinated to the active-site metal in the absence of substrate. This finding has led to a reexamination of earlier X-ray data (Lipscomb et al. 1968) and their reinterpretation has uncovered and confirmed the existence of such a Tyr-248–Zn interaction in the native X-ray crystals (Lipscomb 1973). Second, a conformational change of Tyr-248, induced by interaction of the substrate carboxy group with Arg-145, and proposed solely on the basis of X-ray analysis of the enzyme–Gly-L-Tyr complex, is inconsistent with the findings with esters. The evidence presented by Auld & Holmquist (1974) indicates that esters bind to the metal not to arginine. Hence, the postulated conformational change of Tyr-248 cannot be triggered by an ester–arginine interaction. It must be concluded that esters and peptides are hydrolysed by different catalytic pathways and that neither binds productively to induce the inward movement of Tyr-248.

The organic and inorganic probes that we have employed are all intended to monitor the conformational and structural events coincident with catalysis. Information obtained through the use of such probes provides insight into the dynamics that can then be superimposed on the static picture provided by X-ray crystallography, the only method currently available for three-dimensional structural analysis. The resultant knowledge, when integrated with appropriate functional data, should then allow the deduction of valid mechanisms of action.

ACKNOWLEDGEMENT

This work was supported by Grant-in-Aid GM-15003 from the National Institutes of Health of the US Department of Health, Education and Welfare.

References

ALTER, G. M., LEUSSING, D. L., NEURATH, H. & VALLEE, B. L. (1977) Kinetic properties of carboxypeptidase B in solutions and crystals. *Biochemistry 16,* 3663-3668

ANDERSON, W. F., FLETTERICK, R. J. & STEITZ, T. A. (1974) Structure of yeast hexokinase. III. Low resolution structure of a second crystal form showing a different quaternary structure, heterologous interaction of subunits and substrate binding. *J. Mol. Biol. 86,* 261-269

AULD, D. S. & HOLMQUIST, B. (1974) Carboxypeptidase A. Differences in the mechanisms of ester and peptide hydrolysis. *Biochemistry 13,* 4355-4361

AULD, D. S. & VALLEE, B. L. (1970) Kinetics of carboxypeptidase A. II. Inhibitors of the hydrolysis of oligopeptides. *Biochemistry 9,* 602-609

AULD, D. S., LATT, S. A. & VALLEE, B. L. (1972) An approach to inhibition kinetics. Measurement of enzyme–substrate complexes by electronic energy transfer. *Biochemistry 11,* 4994-4999

BAYNE, S. & OTTESEN, M. (1976) Enzymatically active, cross-linked pig heart lactate dehydrogenase crystals. *Carlsberg Res. Commun. 41,* 211-216

BLAKE, C. C. F. (1976) X-ray enzymology. *FEBS (Fed. Eur. Biochem. Soc.) Lett. 62 (Suppl.),* E30-36

COLEMAN, J. E. & VALLEE, B. L. (1962) Metallocarboxypeptidase–substrate complexes. *Biochemistry 1,* 1083-1092

COLEMAN, J. E. & VALLEE, B. L. (1964) Metallocarboxypeptidase–inhibitor complexes. *Biochemistry 3,* 1874-1879

FÖRSTER, T. (1948) Zwischenmolekulare Energiewanderung und Fluoreszenz. *Ann. Phys 2,* 55

FÖRSTER, T. (1965) in *Modern Quantum Chemistry,* Part III (Sinanoglu, O., ed.), p. 93, Academic Press, New York

HARRISON, L. W., AULD, D. S. & VALLEE, B. L. (1975) Intramolecular arsanilazotyrosine-248·Zn complex of carboxypeptidase A: a monitor of multiple conformational states in solution. *Proc. Natl. Acad. Sci. U.S.A. 72,* 4356-4360

HOLMQUIST, B. & VALLEE, B. L. (1974) Metal substitutions and inhibition of thermolysin: spectra of the cobalt enzyme. *J. Biol. Chem. 249,* 4601-4607

HORROCKS, W. DEW., HOLMQUIST, B. & VALLEE, B. L. (1975) Energy transfer between terbium(III) and cobalt(II) in thermolysin: a new class of metal–metal distance probes. *Proc. Natl. Acad. Sci. U.S.A. 72,* 4764-4768

JOHANSEN, J. T. & VALLEE, B. L. (1971) Differences between the conformation of arsanilazotyrosine 248 of carboxypeptidase A in the crystalline state and in solution. *Proc. Natl. Acad. Sci. U.S.A. 68,* 2532-2535

JOHANSEN, J. T. & VALLEE, B. L. (1973) Conformations of arsanilazotyrosine-248 carboxypeptidase $A_{\alpha, \beta, \gamma}$: comparison of crystals and solution. *Proc. Natl. Acad. Sci. U.S.A. 70,* 2006-2010

JOHANSEN, J. T. & VALLEE, B. L. (1975) Environment and conformation dependent sensitivity of the arsanilazotyrosine-248 carboxypeptidase A chromophore. *Biochemistry 14,* 649-660

JOHANSEN, J. T., KLYOSOV, A. A. & VALLEE, B. L. (1976) Circular dichroism—inhibitor titrations of arsanilazotyrosine-248 carboxypeptidase A. *Biochemistry 15,* 296-303

KANG, E. P., STORM, C. B. & CARSON, F. W. (1975) Cobalt(III) carboxypeptidase A. *J. Am. Chem. Soc. 97,* 6723-6728

KASVINSKY, P. J. & MADSEN, N. B. (1976) Activity of glycogen phosphorylase in the crystalline state. *J. Biol. Chem. 251,* 6852-6859

LATT, S. A. & VALLEE, B. L. (1971) Spectral properties of cobalt carboxypeptidase. The effects of substrates in inhibitors. *Biochemistry 10,* 4263-4269

LATT, S. A., AULD, D. S. & VALLEE, B. L. (1970) Surveyor substrates: energy-transfer gauges of active center topography during catalysis. *Proc. Natl. Acad. Sci. U.S.A. 67*, 1383-1389

LATT, S. A., AULD, D. S. & VALLEE, B. L. (1972) Distance measurements at the active site of carboxypeptidase A during catalysis. *Biochemistry 11*, 3015-3022

LINDERSTRØM-LANG, K. & SCHELLMAN, J. A. (1959) *The Enzymes*, 2nd edn. (Boyer, P. D., Lardy, H. & Myrbäck, K., eds.), vol. 1, pp. 443-510, Academic Press, New York

LINDSKOG, S. & NYMAN, P. O. (1964) Metal-binding properties of human erythrocyte carbonic anhydrase. *Biochim. Biophys. Acta 85*, 462-474

LIPSCOMB, W. N. (1973) Enzymatic activities of carboxypeptidase A's in solutions and crystals. *Proc. Natl. Acad. Sci. U.S.A. 70*, 3797-3801

LIPSCOMB, W. N., HARTSUCK, J. A., REEKE, G.N., QUIOCHO, F. A., BETHGE, P. H., LUDWIG, M. L., STEITZ, T. A., MUIRHEAD, H. & COPPOLA, J. C. (1968) The structure of carboxypeptidase A. VII. The 2.0 Å resolution studies of the enzyme and of its complex with glycyltyrosine, and mechanistic deductions. *Brookhaven Symp. Biol. 21*, 24-90

LUMRY, R. & SMITH, E. L. (1955) The chemical kinetics of some reactions catalysed by pancreatic carboxypeptidase. *Disc. Farad. Soc. 20*, 105-114

NAVON, G., SHULMAN, R. G., WYLUDA, B. J. & YAMANE, T. (1970) Nuclear magnetic resonance study of the binding of fluoride ions to carboxypeptidase A. *J. Mol. Biol. 51*, 15-30

QUIOCHO, F. A. & RICHARDS, F. M. (1966) The enzymic behaviour of carboxypeptidase-A in the solid state. *Biochemistry 5*, 4062-4076

QUIOCHO, F. A., McMURRAY, C. H. & LIPSCOMB, W. N. (1972) Similarities between the conformation of arsanilazotyrosine-248 of carboxypeptidase A in the crystalline state and in solution. *Proc. Natl. Acad. Sci. U.S.A. 69*, 2850-2854

RIORDAN, J. F. (1973) Functional arginyl residues in carboxypeptidase A. Modification with butanedione. *Biochemistry 12*, 3915-3923

RIORDAN, J. F. & MUSZYŃSKA, G. (1974) Differences between the conformations of nitrotyrosyl-248 carboxypeptidase A in the crystalline state and in solution. *Biochem. Biophys. Res. Commun. 57*, 447-451

RIORDAN, J. F. & VALLEE, B. L. (1963) Acetylcarboxypeptidase. *Biochemistry 2*, 1460-1468

ROSSI, G. L. & BERNHARD, S. A. (1970) Are the structure and function of an enzyme the same in aqueous solution and in the wet crystal? *J. Mol. Biol. 49*, 85-91

SAWYER, L. (1972) A fourth crystal form of rabbit muscle aldolase. *J. Mol. Biol. 71*, 503-505

SCHEULE, R. K., VAN WART, H. E., VALLEE, B. L. & SCHERAGA, H. A. (1977) Resonance Raman spectroscopy of arsanilazocarboxypeptidase A: determination of the nature of the azotyrosyl-248·Zn complex. *Proc. Natl. Acad. Sci. U.S.A. 74*, 3273-3277

SIMPSON, R. T. & VALLEE, B. L. (1968) Two differentiable classes of metal atoms in alkaline phosphatase of *Escherichia coli. Biochemistry 7*, 4343-4350

SMITH, E. L. (1949) The mode of action of the metal-peptidases. *Proc. Natl. Acad. Sci. U.S.A. 35*, 80-90

SPILBURG, C. A., BETHUNE, J. L. & VALLEE, B. L. (1974) The physical state dependence of carboxypeptidase A_α as A_γ kinetics. *Proc. Natl. Acad. Sci. U.S.A. 71*, 3922-3926

SPILBURG, C. A., BETHUNE, J. L. & VALLEE, B. L. (1977) Kinetic properties of crystalline enzymes. Carboxypeptidase A. *Biochemistry 16*, 1142-1150

SYTKOWSKI, A. J. & VALLEE, B. L. (1976) Chemical reactivity of catalytic and non-catalytic zinc or cobalt atoms of horse liver alcohol dehydrogenase: differentiation of their thermodynamic and kinetic properties. *Proc. Natl. Acad. Sci. U.S.A. 73*, 344-348

VALLEE, B. L. & RIORDAN, J. F. (1968) Chemical approaches to the mode of action of carboxypeptidase A. *Brookhaven Symp. Biol. 21*, 91-119

VALLEE, B. L. & WILLIAMS, R. J. P. (1968) Metalloenzymes: the entatic nature of their active sites. *Proc. Natl. Acad. Sci. U.S.A. 59*, 498-505

VALLEE, B. L. & LATT, S. A. (1970) in *Structure–Function Relationships of Proteolytic Enzymes* (Desnuelle, P., Neurath, H. & Ottesen, M., eds.), pp. 144-159, Academic Press, New York

VALLEE, B. L., RIORDAN, J. F. & COLEMAN, J. E. (1963) A model for substrate binding and kinetics of carboxypeptidase A. *Proc. Natl. Acad. Sci. U.S.A. 49*, 109-116

VALLEE, B. L., RIORDAN, J. F., BETHUNE, J. L., COOMBS, T. L., AULD, D. S. & SOKOLOVSKY, M. (1968) A model of substrate binding and kinetics of carboxypeptidase A. *Biochemistry 7*, 3547-3556

VAN WART, H. E. & VALLEE, B. L. (1977) Exchange inert Co(III)-carboxypeptidase A; a catalytically inactive metallocarboxypeptidase. *Biochem. Biophys. Res. Commun. 75*, 732-738

WALDSCHMIDT-LEITZ, E. (1931) The mode of action and differentiation of proteolytic enzymes. *Physiol. Rev. 11*, 358-370

WEBER, G. (1975) Energetics of ligand binding to proteins. *Adv. Protein Chem. 29*, 1-83

WELLS, M. A. & BRUICE, T. C. (1977) Intramolecular catalysis of ester hydrolysis by metal complexed hydroxide ion. Acyl oxygen bond scission in Co^{2+} and Ni^{2+} carboxylic acid complexes. *J. Am. Chem. Soc. 99*, 5341-5356

Discussion

Battersby: On what scale do you work?

Vallee: Very small; µg or sometimes mg are sufficient. When one wants to measure distances, the stop-flow experiments are usually done with equimolar enzyme and substrate, but otherwise we use catalytic amounts of enzymes. The resonance Raman methods do not require unusual amounts of material. The important point which I should stress is that the chemical modifications have to be absolutely defined; each has to be checked to ensure chemical stoichiometry and specificity. If and when problems are encountered, this is where they are encountered.

Lipscomb: One benefit of the arsanilazotyrosine probe is that it is active as a peptidase. Most of the other carboxypeptidases in which Tyr-248 is modified are not.

Vallee: It is active both as a peptidase and esterase; it is a fully active enzyme. We are puzzled, however, by the fact that we can modify Tyr-248 in various ways and alter the peptidase activity drastically—in a few cases to the point of disappearance. Esterase activity, however, not only remains intact but is usually augmented. The two azo probes—the tetrazolylazo and the arsanilazo one—give different results, even though they modify the same residue. Thus, the chemical characteristics of Tyr-248 depend tremendously on the nature of the modification and Professor Lipscomb agrees that it is not a steric problem.

Lipscomb: The reagent would fit in the molecule in the crystal.

Topping: Did you try many probes before you settled on that one and how did you choose it?

Vallee: We began to examine a series of azo dyes of the sort studied by Tabachnik & Sobotka (1959, 1960). Most of these agents also modify lysine, histidine, and cysteine besides tyrosine. We wanted to exploit the circular dichroic spectral properties of the derivatives which might be environmentally-sensitive probes of protein structure. We chose diazotized arsanilic acid because we wanted an independent measure of incorporation into the protein—not only could we measure the spectrum and the amino acid composition but the arsenic content as well.

Williams: Just what steps are you following in this process? Is the step after the initial binding a rearrangement of the substrate in the pocket or formation of an intermediate on the reaction path?

Vallee: There is a finite number of possibilities—Professor Lipscomb has covered many of them and we have covered some. One important factor is that the mechanisms of ester and peptide hydrolysis for this enzyme are simply not the same. At one stage it was thought that Co(III) might help us in this regard but this has not proven to be so. We do not know whether there is an anhydride intermediate or even whether there is general base catalysis. Until such time that we can, for example, slow the reaction down sufficiently to see the intermediate in detail, we cannot answer your question. The glutamic acid is an important group—that is clear from all available evidence—but whatever else might be participating is unknown. It seems to me that the only approach is that which you described—a crystallographic and structural approach combined with methods that have resolving times short enough for this purpose, and with knowledge of the residue with which we are dealing. That we are trying to do.

Lipscomb: We studied the binding of glycyltyrosine by X-ray crystallography. Because it is bound in the pocket, in the position in which we find most inhibitors in several other studies, we conclude that it is largely in the productive mode of binding. However, there appears to be an anomaly in the binding of glycyltyrosine which renders the rate of cleavage about 2000 times less rapid than that of longer substrates. The NH_2-terminus of glycyltyrosine is linked through a water molecule by a hydrogen-bond to Glu-270 which is supposed either to promote the attack of water or to be the nucleophile which makes an anhydride. We cannot build long substrates with that conformation; to do so, we must make a rotation about a single bond so that the NH_2 group is oriented in a different direction. That rotation would set Glu-270 free to fulfil its function. Hence, a conformational change is required in the case of glycyltyrosine and perhaps for even longer substrates as well.

Vallee: I must emphasize that the only substrate for which one has structural information is glycyltyrosine and, for the reasons just indicated, it is difficult to know if we can extrapolate to longer substrates. Any study of the mechanism must be based on what is physically seen at the moment of catalysis. Anything else is sheer conjecture. We do know that minimally two conformations of the protein are induced by these substrates and the probability is high for such conformational changes with other substrates, as well.

Franks: P. Douzou & I have used liquid supercooled water as solvent for kinetic studies of enzymes. This has obvious advantages over the commonly-used aqueous-organic mixtures. In this way we can stabilize otherwise-labile intermediates almost indefinitely (Douzou *et al.* 1978).

Vallee: Auld has designed a stopped-flow spectrophotometer for use at low temperatures to obtain spectra of carboxypeptidase during catalysis and we plan to do experiments analogous to those you just suggested.

Franks: Do you use aqueous-organic mixtures?

Vallee: Sometimes, but not always.

Franks: With supercooled water the system can be kept liquid down to almost −40 °C, at which temperature many enzyme reactions come to a standstill for practical purposes.

Vallee: Our technique can be used in other ways; we did not design our instrument solely for that purpose.

Arigoni: Have the stereoelectronic arguments developed by Deslongchamps (1975) for simple model systems been applied to the reaction catalysed by carboxypeptidase, especially in view of this duality of activity towards esters and peptides, much along the lines that Professor Dunitz referred to previously, namely that developed by Bizzozero & Zweifel (1975) for chymotrypsin?

Vallee: Many attempts have been made, by Breslow & Wernick (1977), by Makinen *et al.* (1976) and by others. The question is whether any model that does not incorporate binding and catalysis in a manner analogous to that just described will suffice.

Arigoni: But the Deslongchamps approach allows one to make some accurate predictions about what stereochemical element ought to be present for the reaction to occur. It works well *in vitro* and we should like to know how far these ideas can be extended to the *in vivo* case.

Vallee: I should be delighted if somebody succeeded. What bothers me is that the observed conformational changes, induced in the manner I described, were unpredictable. How critical is the juxtaposition of those residues? How must they be aligned at the moment of catalysis to operate in a concerted manner? There are few instances of any kind in which reaction mechanisms have been established definitely, let alone with enzymes.

Arigoni: But one can never prove mechanisms; one can only rule out alternatives.

Vallee: That is so, but one can find instances in which everything is consistent with what one has proposed.

Lipscomb: The enzyme thermolysin has an active site somewhat like that in carboxypeptidase but it cleaves peptides in the middle of the chain. Matthews suggests that the glutamic acid is too remote from the substrate to act directly to form an anhydride. He sees an intervening water molecule in the inhibitor complexes. That result tends to support the hypothesis that, for thermolysin, glutamic acid promotes the attack of water as a nucleophile. The putative proton donor is histidine, not tyrosine. Otherwise these active sites are much the same.

Vallee: Kester & Matthews (1977) have compared these two active sites and concluded that they have several features in common but are not identical.

[*See also pp. 265-269.*]

References

BIZZOZERO, S. A. & ZWEIFEL, B. O. (1975) The importance of the conformation of the tetrahedral intermediate for the chymotrypsin-catalyzed hydrolysis of peptide substrates. *FEBS (Fed. Eur. Biochem. Soc.) Lett. 59,* 105-107

BRESLOW, R. & WERNICK, D. L. (1977) Unified picture of mechanisms of catalysis of carboxypeptidase A. *Proc. Natl. Acad. Sci. U.S.A. 74,* 1303-1307

DESLONGCHAMPS, P. (1975) Stereochemical control in the cleavage of tetrahedral intermediates in the hydrolysis of esters and amides. *Tetrahedron 31,* 2463

DOUZOU, P., DEBEY, P. & FRANKS, F. (1978) Supercooled water as medium for enzyme reactions at subzero temperatures. *Biochim. Biophys. Acta 523,* 1-8

KESTER, N. R. & MATTHEWS, B. W. (1977) Comparison of the structures of carboxypeptidase A and thermolysin. *J. Biol. Chem. 252,* 7704-7710

MAKINEN, M. N., YAMAMURA, K. & KAISER, E. I. (1976) Mechanisms of action of carboxypeptidase A in ester hydrolysis. *Proc. Natl. Acad. Sci. U.S.A. 73,* 3882-3886

TABACHNIK, M. & SOBOTKA, H. (1959) Azoproteins I. Spectrophotometric study of amino acid azo derivatives. *J. Biol. Chem. 234,* 1726-1730

TABACHNIK, M. & SOBOTKA, H. (1960) Azoproteins II. A spectrophotometric study of the coupling of diazotized arsanilic acid with proteins. *J. Biol. Chem. 235,* 1051-1054

To stabilize a transition state . . .

JOSHUA BOGER and JEREMY R. KNOWLES

Department of Chemistry, Harvard University, Cambridge, Massachusetts

Abstract   Inspection of the active sites of the many enzymes whose structures are known at high resolution leads to the unsurprising conclusion that an enzyme may provide an environment that exquisitely stabilizes the transition state for an elementary catalytic step that is expected to be difficult. In an effort both to mimic such an environment and to have the opportunity of investigating the thermodynamic and kinetic consequences of juxtaposing polar and non-polar loci in the same molecule, we have synthesized several specifically functionalized α-cyclodextrins. One of these is designed to stabilize the trigonal bipyramidal transition state for an in-line displacement at the phosphorus of a phosphate monoester. This cyclodextrin contains three *symmetrically disposed* ammonium groups on the 'top' (at C-6) of the hydrophobic cavity formed by the hexa-glucose torus and the remaining 15 hydroxy groups are methylated. The thermodynamic consequences of adjacent hydrophobic and hydrogen-bonding and/or electrostatic binding sites are investigated using several charged and uncharged ligands. The feasibility of building host species explicitly to stabilize reaction transition states is discussed.

Over the past 20 years or so, physical organic enzymologists have taken two different attitudes in their attempts to account for the effectiveness of enzymes as catalysts. The first position has been to define a set of more-or-less identifiable features that may together produce the kind of rate acceleration that we can estimate by comparing an enzyme with a simple organic catalyst that effects the same (or a similar) chemical conversion. Such features as approximation, orientation, strain, 'push-pull' catalysis, multiple reaction intermediates, and the 'micro-environment' have all been analysed, and model systems that exemplify these features have been explored. This activity has led to a rich literature on intramolecular reactions and neighbouring-group participation, on general acid-base and nucleophilic catalysis, and on solvent effects (see, e.g., Jencks 1975; Guthrie 1976). Yet although we may be satisfied that these features could together provide more than enough rate enhancement to account for the most efficient enzymes, the approach has not led to such a dramatic advance in our understanding that we can design a catalyst *de novo*.

The second approach owes more to the detailed study of enzymes themselves, and particularly to the structural picture of productive enzyme–substrate complexes that has been increasingly refined for several enzymes over the past decade (e.g., Kraut 1977). What do we learn from looking at these structures? In the case of such well studied enzymes as the serine proteinases or lyzome, we see how the substrate is held at the active site by several binding interactions and how the constellation of functional groups around the bonds being formed and broken are arranged *so as exquisitely to stabilize the transition state for an elementary step that we expect to be difficult in the overall reaction.* This stabilization, coupled with the possibility of unusual chemical reactivity of particular groups at the active site, provides both an attractive view of enzyme catalysis, and opens up a more obvious approach to its emulation.

At the simplest level, we aim to create a catalytic species for a one-step reaction by providing an environment that is complementary in structure and electron distribution to the transition state for the reaction. The aim may also be fulfilled by making an environment in which the substrate is distorted structurally and electronically towards the transition state. (The enzyme literature is replete with discussions on the difference between these two statements, but in terms of both catalyst design and of the kinetic consequences of substrate interacting with a catalyst, the difference is largely semantic). Further, we want to investigate a model system in which the elements of catalysis may be more readily discerned and more precisely defined, rather than simply to search for a host species that provides a large but still inexplicable rate enhancement.

In order to stabilize a transition state in preference to the starting material or the product, a system must be chosen that satisfies the following criteria:

(*a*) There should be a substantial geometrical change both as the substrate reaches the transition state and as the transition state collapses to form the product. The larger this change in geometry, the easier it will be to create a host species that is complementary to a transition state whose structure we can in any case only guess at.

(*b*) The transition state should contain structural features that provide 'handles' for stabilizing interactions (e.g., developing charges that may be delocalized, hydrogen-bond donors that can be provided with acceptors, etc.).

(*c*) The catalyst species should be a reasonably rigid molecule. A 'floppy' host would presumably accommodate substrate, transition state and product indiscriminately, which would result in binding but no catalysis.

(d) It will be desirable to design into the system some form of fine tuning, so that the precise location of substrate with respect to the catalytic groups of the host can be adjusted, and the required distinction between ground state and transition state can be maximized. We must, therefore, be able to make structural adjustments in both substrate and catalyst, with minimal effort.

(e) Finally, the major source of the interaction free energy between substrate and catalytic host must be different from the interactions that stabilize the transition state at the reaction centre (b, above). In enzymological language, the substrate-binding free energy must derive (in part, at least) from regions of the substrate unconnected with the reaction centre. This separation between binding loci and catalytic groups is crucial, for it allows binding free energy to be used catalytically. The binding locus will position the substrate's reaction centre with respect to the catalytic locus and the groups at the catalytic locus will ensure that it is the transition-state geometry that is preferentially stabilized. The importance of this separation, which has been persuasively argued (if not quite in these terms) by Breslow (1969), Jencks (1975), and others, will become more clear from what follows.

THE CHOICE OF REACTION

One of the reactions that seems to fit the above criteria nicely is the solvolysis of a phosphate monoester (see, e.g., Benkovic & Schray 1973). A considerable body of evidence suggests that the favoured pathway for the hydrolysis of phosphate monoester monoanions proceeds *via* monomeric metaphosphate. If the leaving group alcohol has a low pK_a, the ester dianions may also fragment directly. As the leaving group becomes less good, the monoanions start to be hydrolysed faster than the dianions, possibly *via* a pre-equilibrium proton transfer that protonates the leaving group oxygen. For these dissociative mechanisms involving metaphosphate, therefore, there is a transfer of charge from the phosphoryl oxygen atoms to the leaving group, which may itself require protonation in the case of weakly acidic alcohols. However, it is difficult to formulate catalytic mechanisms that would accelerate the formation of metaphosphate, other than by lowering the effective pK_a of the leaving group.

Consider, on the other hand, the frequent pathway of hydrolysis of phosphate di- and tri-esters, which involves a simple 'in-line' displacement *via* a transition state that is pentacoordinate at phosphorus. This pathway has not been demonstrated conclusively for a phosphate monoester, but its attractiveness is based on the view that the trigonal-bipyramidal transition state *is* amenable to stabilization and, therefore, to more facile acceleration by an appropriate catalyst. As mentioned above, it is difficult to see how phosphoryl group transfer could be catalysed easily if the metaphosphate pathway is followed, and the conventional wisdom is that enzyme-catalysed transfers of phosphoryl groups proceed *via* associative in-line displacements. (Thus: '... groups capable of serving as rigid binding sites may be expected in positions consistent with the geometry of pentacovalent phosphorus.' [Benkovic & Schray 1973]). This is the single-step process we have chosen.

THE CHOICE OF CATALYTIC HOST

Consider the hydrolysis of monobenzyl phosphate. According to the criteria outlined, we want preferentially to stabilize the trigonal-bipyramidal transition state for an in-line attack at phosphorus, and this can in principle be achieved (i) by providing three electrophilic centres appropriately positioned to interact with and stabilize the partially anionic oxygen atoms, and (ii) by providing a binding locus for the benzyl group such that trigonal-bipyramidal rather than tetrahedral geometry is forced upon phosphorus (see Fig. 1). On the basis of simple model-building (using space-filling models) a promising candidate for the catalyst required in Fig. 1C is the symmetrical triamino α-cyclodextrin shown in Fig. 3 [6,6'',6''''''-triamino-6,6'',6''''''-trideoxy-6',6''',6'''''-tri-*O*-methyl-2,2',2'',2''',2'''',2''''',3,3',3'',3''',3'''',3'''''-dodeca-*O*-methyl-α-cyclodextrin≡symmetrical triamino-per-*O*-methyl-α-cyclodextrin].

The cyclodextrins are a family of cyclic α-(1-4)-linked oligoglucoses formed from starch by *Bacillus macerans* (French 1957). The smallest member of the family is α-cyclodextrin, which contains six D-glucose residues that form a toroid 0.9 nm (9 Å) high and 1.35 nm (13.5 Å) in diameter. The cavity is hydrophobic and has an internal diameter of about 5.5 Å (Fig. 2). At the 'top' of the torus are the six primary hydroxy groups (on C-6), and the 'bottom' is ringed by the twelve secondary hydroxy groups (on C-2 and C-3). Cyclodextrins bind a wide variety of non-polar ligands with dissociation constants in the mmol/l range (Cramer *et al.* 1967; Saenger 1976; Bergeron *et al.* 1977; Wood *et al.* 1977; Behr & Lehn 1976). The variation in dissociation constants is not dramatic, except for a few compounds whose bulk sterically prevents their inclusion into the cavity. The energetic basis of this binding has been ascribed to release of conformational strain in the cyclodextrin (Manor & Saenger 1974), to London dispersion forces (Bergeron *et al.* 1977), and to hydrophobic interactions (Griffiths & Bender 1973). For present purposes, however, the existence of an interaction free energy for a single aromatic ring

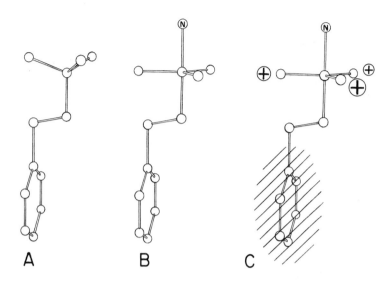

FIG. 1. (A) Benzyl phosphate; (B) transition state for in-line attack of a nucleophile N on benzyl phosphate; (C) required disposition of binding and catalytic groups for the catalysis of an in-line displacement. The symbols ⊕ represent three electrophilic groups that delocalize the partial charges on the equatorial oxygen atoms; the shaded area is a binding locus that both provides the favourable catalyst–substrate interaction, and forces trigonal-bipyramidal geometry on phosphorus by pushing phosphorus into the plane of the electrophiles.

of around 16–20 kJ/mol suffices. This binding energy (which is in the range of that for many enzyme–substrate complexes), coupled with the possibility of chemically modifying the hydroxy groups, has led to several studies directed at modelling enzyme catalysis (Breslow & Overman 1970; Breslow 1971; Cramer & Helter 1967; Griffiths & Bender 1973; Mochida *et al.* 1976; Siegel *et al.* 1977).

In this paper we report the synthesis of the symmetrical triamino-per-*O*-methyl-α-cyclodextrin (Fig. 3), designed to contain the basic features required for a catalyst for a displacement reaction at phosphorus. The major interaction free energy between an aryl phosphate ester and the catalyst derives from the interaction between the aryl group and the cyclodextrin cavity. The three ammonium ions should provide electrostatic (and/or hydrogen-bonding) delocalization of the charge(s) on the phosphoryl group oxygen atoms. The remaining 15 hydroxy groups are masked by methylation. The host molecule is reasonably rigid, and by varying the distance between the hydrophobic part of the substrate and the phosphorus atom (for instance, changing benzyl phosphate to phenyl phosphate), or by varying the substrate substituents, we should be able to 'tune' the system so that phosphorus is forced from

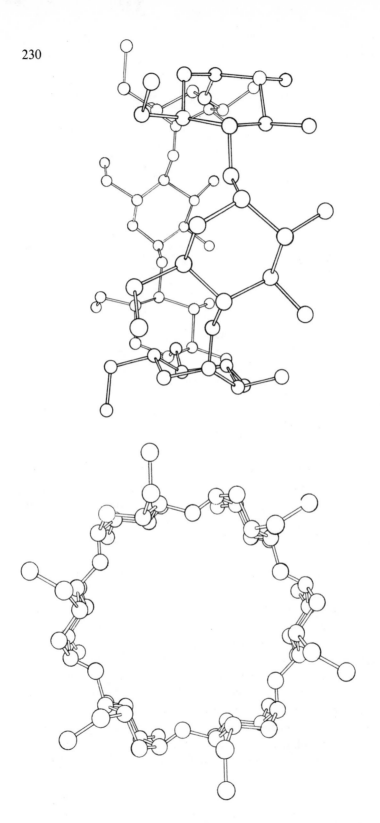

FIG. 2. Drawings of unmodified α-cyclodextrin. The coordinates used for the cyclodextrin skeleton were based on crystal structure data summarized by Saenger (1976): left, looking down on the primary (C-6) rim; right, side view.

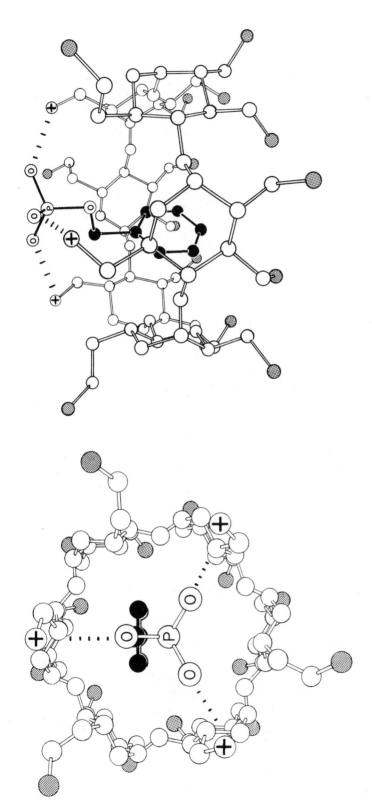

FIG. 3. Drawings of the symmetrical triamino-per-O-methyl-α-cyclodextrin complex with benzyl phosphate. Each of the three ammonium ions is shown by the symbol ⊕. The 15 methyl groups are shaded. All hydrogen atoms are omitted. The size of the atoms is arbitrary.

tetrahedral geometry towards a trigonal bipyramid. We expect that the addition either of large hydrophobic groups (e.g. t-butyl, phenyl) or of polar groups (e.g. ammonium) at the *m*- or *p*-position of the substrate will result in predictable changes in the positioning of the substrate with respect to the 'catalytic' groups on the cyclodextrin. Model-building suggests that the distances between the ammonium ions and the phosphoryl oxygen atoms of a bound benzyl phosphate are in the range required for strong hydrogen bonds (2–3 Å). When the aromatic ring is positioned optimally within the cavity, it seems probable that some flattening of the tetrahedral disposition of phosphoryl oxygen atoms will occur (Fig. 3). To make changes of this kind rationally we shall need information about the structures and energetics of such complexes. Our steps in the acquisition of this information are outlined below.

SYNTHESIS OF THE MODIFIED CYCLODEXTRIN

To modify three out of the 18 hydroxy groups of α-cyclodextrin we need two levels of discrimination. First, we must modify only the *primary* groups (at C-6) and secondly we must isolate only the *symmetrical* species from a mixture of the (four) tris-modified cyclodextrins. Our synthetic route is shown in Fig. 4. Trityl chloride is quite selective for the primary hydroxy groups and in mild conditions allows the isolation of di-, tri- and tetra-substituted species (there are in principle 10 different kinds of molecule in this mixture!). By medium-pressure liquid ('short-column') chromatography on silica gel (Hunt & Rigby 1967), we have isolated the symmetrical tritrityl material in about 20% yield from α-cyclodextrin. The characterization of this species is based primarily on the extensive use of ^{13}C n.m.r. spectroscopy, which allows an unequivocal discrimination amongst the symmetrical and unsymmetrical tri-substituted compounds to be made. After *O*-methylation of the remaining 15 hydroxy groups, hydrolysis of the trityl groups, mesylation and displacement with azide ion, the symmetrical triazido-per-*O*-methyl-α-cyclodextrin was reduced to the triamino-per-*O*-methyl compound. These last five steps in the synthetic scheme (Fig. 4) have been perfected so that the yield of the triamino-per-*O*-methyl compound from the tritrityl derivative is essentially quantitative (> 95%). Although the full characterization of the final product will not be given here, Fig. 5 shows the ^{13}C n.m.r. spectrum of the symmetrical tritrityl-α-cyclodextrin, and Fig. 6 shows the ^{1}H n.m.r. spectrum of the tritrityl-per-*O*-methyl derivative. The legends to these figures contain some of the assignments, and particularly informative regions from spectra of an *un*symmetrical analogue are included as insets in these spectra to convince the reader that the

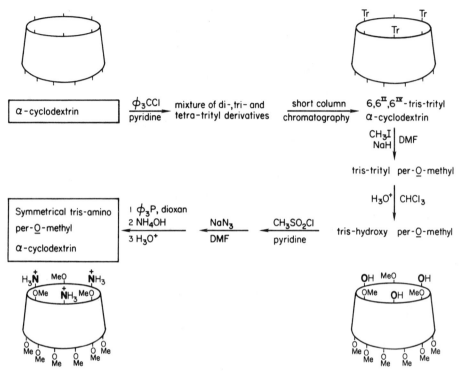

FIG. 4. Synthetic scheme for symmetrical triamino-per-*O*-methyl-α-cyclodextrin.

correct isomer has been isolated. The monoamino-per-*O*-methyl derivative was prepared by a similar procedure.

LIGAND BINDING TO CATIONIC α-CYCLODEXTRINS

Before attempting any kinetic experiments, we must first establish the ligand-binding properties of these bifunctional host species. Not only must the pK_a values for the ionizable groups be known, but the preferred binding modes of simple aromatic compounds must be determined, so that productive (i.e., 'right-way-round') substrate binding may be assured.

The titration curve for triamino-per-*O*-methyl-α-cyclodextrin is shown in Fig. 7. This curve precisely yields the three pK_a values which differ from one another by somewhat more than the 0.477 units expected merely on statistical grounds. Evidently the three amino groups do feel each other's presence to some extent, even though the effect is relatively modest. These results also provide the important practical information about the ionization state of the uncomplexed cyclodextrin at particular pH values.

TABLE 1

Dissociation constants of various ligands from modified and unmodified α-cyclodextrins (α-CD)[a]

Ligand	α-Cyclodextrin	Monohydroxy-per-O-methyl-α-CD	Monoamino-per-O-methyl-α-CD	Symmetrical trihydroxy-per-O-methyl-α-CD	Symmetrical triamino-per-O-methyl-α-CD
4-Nitrophenol (pH 2)[b]	2.5 (0.4)	0.71 (0.11)	1.2 (0.3)	1.2 (0.1)	2.5 (0.3)
4-Nitrophenyl phosphate (pH 7)[c]	8.1 (1.6)	No binding[d] observed	27 (5)	30 (2)	1.1 (0.2)

[a] Determined spectrophotometrically, I = 0.1, 30 °C. All dissociation constants are in mmol/l, obtained by a modificatiom of the method of Hildebrand & Benesi (1949) (see Rossotti & Rossoti 1961). The numbers in parentheses are the standard deviations derived from several linear plots at widely separated wavelengths.

[b] HCl/KCl.

[c] Triethanolamine–HCl buffer.

[d] i.e., $K_{diss.}$ > 100 mmol/l.

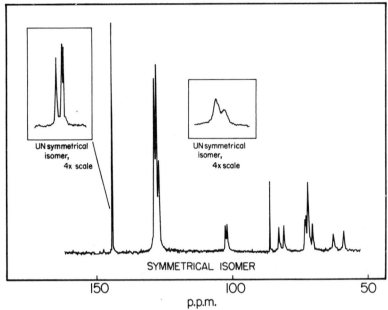

FIG. 5. ^{13}C n.m.r. spectrum of symmetrical trityl-α-cyclodextrin. The spectrum was recorded at 25.2 MHz, in solution of [^{2}H$_6$]Me$_2$SO. The peak near 145 p.p.m. is a singlet, showing only one kind of *ipso*-trityl group carbon atom, and the peak near 105 p.p.m. is a doublet, showing only two kinds of C-1 glucose carbon atoms, in the symmetrical molecule. *Insets:* corresponding regions for an *un*symmetrical trityl derivative. Left, three different *ipso*-carbon atom environments are seen; right, a multiplicity (theoretically, six) of different C-1 carbon environments are seen.

The binding of ligands (that are more substantial than protons) to cyclodextrins has been studied by several methods, the most convenient of which involves the measurement of the change in the ultraviolet–visible spectrum of chromophoric substrates on binding in the cyclodextrin cavity. This is the method we have used. It must be emphasized that a small dissociation constant does not necessarily mean that the ligand is fully included in the cyclodextrin, and other evidence is required both to confirm inclusion and to specify which way round the ligand binds.

Table 1 lists the results of some preliminary experiments designed to establish (*a*) the qualitative importance of the electrostatic interaction, if any, and (*b*) the sense of binding (i.e., which way round the anionic ligands are included). First, it is evident that 4-nitrophenol binds more tightly to the cyclodextrin when most of the hydroxy groups have been methylated. The dissociation constants of the complexes with the monohydroxy-permethyl and monoamino-permethyl-cyclodextrin are significantly lower than with α-cyclodextrin itself. Evidently the increase in the hydrophobic character of the

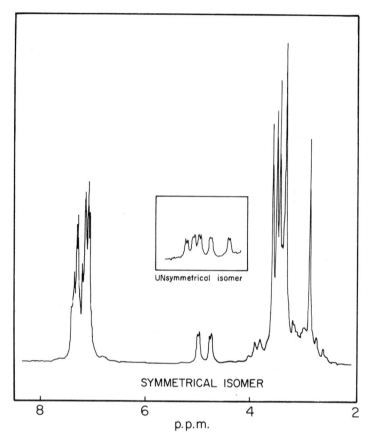

UNsymmetrical isomer

SYMMETRICAL ISOMER

8   6   4   2

p.p.m.

FIG. 6. ^{1}H n.m.r. spectrum of symmetrical trityl-per-O-α-cyclodextrin. The spectrum was recorded at 100 MHz, in C^{2}HCl$_3$ solution. The pair of doublets near 4.8 p.p.m. derive from the two kinds of C-1 proton (split by the C-2 proton) in the symmetrical trityl compound. *Inset:* the corresponding region for an unsymmetrical trityl derivative. There are in fact, six protons (one doublet is unresolved) in this set of peaks, which show six different C-1 proton environments.

cyclodextrin (or at least in the physical extent of this characteristic) results in tighter binding. There is no doubt from (*i*) the substantial change in extinction coefficient (Δε) of these chromophores on binding and (*ii*) the presence of two isosbestic points in the spectrophotometric titration that the phenol is being included in the cavity, and we observe the expected increase in Δε on going from α-cyclodextrin to the per-O-methyl derivatives. On the basis of molecular models, 4-nitrophenol is not, however, expected to interact especially favourably either with the hydroxy group in monohydroxy-cyclodextrin or with ammonium ion in monoamino-cyclodextrin. Indeed, when the number of functional groups is increased (in the tri-functional cyclodextrins) the bind-

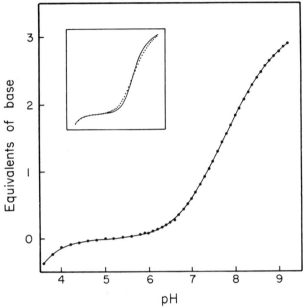

FIG. 7. pH-tritration curve for triamino-per-*O*-α-cyclodextrin. The line is theoretical, for pK_a values of 7.05 ± 0.01, 7.75 ± 0.02, and 8.49 ± 0.01 derived from the experimental points according to Sayce (1968). The titration was performed at 40 °C, I = 0.1. *Inset:* theoretical line showing the poor fit to the experimental points for the titration of three *non-interacting* groups, where the observed pK_a values are separated only by the statistical factor of 0.477, and are 7.27, 7.75, and 8.23.

ing is marginally less strong, a fact which suggests that hydrophobic forces dominate the free energy of binding for this non-specific ligand.

When 4-nitrophenyl phosphate is bound more dramatic changes are seen (Table 1). Although this ligand (which at pH 7 is mostly dianionic) binds satisfactorily to α-cyclodextrin itself, no interaction with the monohydroxy-cyclodextrin is observable. This is not surprising, since the phosphoryl group must occupy a hydrophobic environment in whichever sense the ligand sits in the cavity. Changing the monohydroxy compound into the monoammonium derivative gives a weak but measurable binding and suggests that we are here observing some interaction between the charges. Provision of three appropriately placed hydrogen-bond donors (in the trihydroxy compound) provides about the same stabilization as a single charge. When, however, the triamino-cyclodextrin is used (and this exists largely as the trication at pH 7, see Fig. 7), a sharp increase in binding is seen. Using the data for the neutral ligand 4-nitrophenol as a base, it is evident that the charge–charge interaction between 4-nitrophenyl phosphate and the triamino host lowers the dissociation constant by more than 50-fold. To be certain that this binding of 4-nitrophenyl phosphate is in the expected sense we are testing 3,5-dimethyl-4-nitrophenyl phosphate. This molecule cannot penetrate the cyclodextrin cavity in the 'wrong' sense, but makes a snug fit (according to molecular models) if the phosphate anion(s) interact with the ammonium cation(s) of the triamino host in the 'right' way depicted in Fig. 3.

Meanwhile, a small catalytic effect has been seen for 4-nitrophenyl phosphate and the triamino-per-O-methylcyclodextrin. It remains to be seen whether the more proper positioning of the phosphoryl group with respect to the host's functionalities will lead to the intended kinetic consequences.

ACKNOWLEDGEMENTS

We are indebted to Daniel Brenner for the synthesis of the monoamino-per-O-methyl-cyclodextrin, to John Ramsdale, Steven Abbott, and Joel Belasco for the molecular drawing programme, and to the National Science Foundation for support.

References

BEHR, J. P. & LEHN, J.-M. (1976) Molecular dynamics of α-cyclodextrin inclusion complexes. *J. Am. Chem. Soc. 98*, 1743-1747

BENKOVIC, S. J. & SCHRAY, K. J. (1973) Chemical basis of biological phosphoryl transfer, in *The Enzymes*, 3rd edn., vol. 8 (Boyer, P. D., ed.), pp. 201-239, Academic Press, New York

BERGERON, R. J., CHANNING, M. A., GIBERLY, G. J. & PILLOR, D. M. (1977) Disposition requirements for binding in aqueous solution of polar substrates in the cyclohexaamylose cavity. *J.Am. Chem. Soc. 99*, 5146-5151

BRESLOW, R. (1969) *Organic Reaction Mechanisms,* 2nd edn., pp. 57-66, W. A. Benjamin, New York

BRESLOW, R. (1971) Studies on enzyme models, in *Bioinorganic Chemistry, Adv. Chem. Ser. 100* (Gould, R. F., ed.), pp. 21-43, American Chemical Society, Washington

BRESLOW, R. & OVERMAN, L. E. (1970) An 'artificial enzyme' combining a metal catalytic group and a hydrophobic binding cavity. *J. Am. Chem. Soc. 92*, 1075-1077

CRAMER, F. & HETTLER, H. (1967) Inclusion compounds of cyclodextrins. *Naturwissenschaften 54*, 625-632

CRAMER, F., SAENGER, W. & SPATZ, H.-CH. (1967) Inclusion compounds. XIX. The formation of inclusion compounds of α-cyclodextrin in aqueous solutions. Thermodynamics and kinetics. *J. Am. Chem. Soc. 89*, 14-20

FRENCH, D. (1957) The Schardinger dextrins. *Adv. Carbohydrate Chem. 12*, 189-260

GRIFFITHS, D. W. & BENDER, M. L. (1973) Cycloamyloses as catalysts. *Adv. Catal. 23*, 209-261

GUTHRIE, J. P. (1976) Enzyme models and related topics, in *Techniques of Chemistry*, vol. 10, part 2 (Jones, J. B., Sih, C. J., & Perlman, D., eds.), pp. 628-730, John Wiley, New York

HILDEBRAND, J. H. & BENESI, H. A. (1949) A spectrophotometric investigation of the interaction of iodine with aromatic hydrocarbons. *J. Am. Chem. Soc. 71*, 2703-2707

HUNT, B. J. & RIGBY, W. (1967) Short column chromatography. *Chem. Ind.,* 1868-1869

JENCKS, W. P. (1975) Binding energy, specificity and enzyme catalysis. The Circe effect. *Adv. Enzymol.,* 219-410

KRAUT, J. (1977) Serine proteases: structure and mechanism of catalysis. *Annu. Rev. Biochem. 46*, 331-358

MANOR, P. C. & SAENGER, W. (1974) Topography of cyclodextrin inclusion complexes. III. Crystal and molecular structure of cyclohexaamylose hexahydrates, the $(H_2O)_2$ inclusion complex. *J. Am. Chem. Soc. 96*, 3630-3639

MOCHIDA, K., MATSUI, Y., OTA, Y., ARAKAWA, K. & DATE, Y. (1976) Substrate specificity in the cyclodextrin-catalyzed cleavage of organic phosphates and monothiophosphates in alkaline solutions. *Bull. Chem. Soc. Jpn. 49*, 3119-3123

ROSSOTTI, F. J. C. & ROSSOTTI, H. (1961) *The Determination of Stability Constants*, p. 276, McGraw-Hill, New York

SAENGER, W. (1976) α-Cyclodextrin inclusion complexes, mechanism of adduct formation and intermolecular interactions, in *Environmental Effects on Molecular Structure* (Pullman, B., ed.), pp. 265-305, D. Reidel Publishing Co., Dordrecht, Holland

SAYCE, I. G. (1968) Computer calculation of equilibrium constants of species present in mixtures of metal ions and complexing agents. *Talanta 15*, 1397-1411

SIEGEL, B., PINTER, A. & BRESLOW, R. (1977) Synthesis of cycloheptaamylose 2-, 3-, and 6-phosphoric acids, and a comparative study of their effectiveness as general acid or general base catalysts with bound substrates. *J. Am. Chem. Soc. 99*, 2309-2312

WOOD, D. J., HRUSKA, F. E. & SAENGER, W. (1977) ^1H NMR study of the inclusion of aromatic molecules in α-cyclodextrin. *J. Am. Chem. Soc. 99*, 1735-1740

Discussion

Kenner: At the beginning of the catalytic reaction the benzyloxy anion is countered by the phosphate but how does the anion leave the cyclodextrin? It sits in a hydrophobic hole surrounded by methylated groups with no room even for water to get in.

Knowles: The leaving group needs a proton donor and we have not provided one. But, unless there is considerable development of charge on the benzyl oxygen atom at the transition state, we need not be concerned. We need not worry about what happens after the system has surmounted the free energy barrier. That is, unless we have been so extraordinarily successful as to make a free-energy *well* for the transition state!

Dunitz: Obviously it will be easier to observe stabilization for a transition state that is not too high in energy. For example, stabilization of a transition state of, say, 100 kcal/mol by 10 kcal/mol will have little effect on the rate whereas similar stabilization of a 25–35 kcal/mol transition state will be noticeable.

Knowles: If the activation free energy is decreased by 10 kcal/mol, that means a rate acceleration of 10^7.

Dunitz: That hardly matters if the rate is 10^{-50} s^{-1}.

Knowles: I agree that any catalytic acceleration that we observe will be a lower limit, because we shall have changed the mechanistic pathway (from a dissociative metaphosphate route to an associative in-line displacement). But the half-life of, say, *p*-nitrophenyl phosphate in aqueous buffer at pH 7 and 30 °C is about one year—it is not immeasurably small. That of *p*-nitrobenzyl phosphate may be somewhat longer. But we may change the solvent, ionic strength and pH (as well as the substrate) so as to change the contributions of the electrostatic and hydrophobic interactions, and optimize the catalysis. I must confess that we were delighted to detect an electrostatic-binding interaction in aqueous solution at all.

Topping: Perhaps the effect on your observed rate constants of a change in ionic strength of the reaction medium might be even more revealing than a change in solvent. As the expected reaction involves interaction between a dianion and a trication in the transition state, a Brønsted–Bjerrum plot showing a slope of −6 would be convincing evidence for such a mechanism.

Knowles: Yes, indeed; that would be nicely diagnostic.

Williams: May I give a similar but simpler example of how to devise the chemistry for catalysis and manufacture the catalyst? We have looked at simple unimolecular-type reactions which handle electrons. The transition state we wanted to build was that for electron transfer and so posed no volume problem but the same type of geometric factor problem as yours. For the catalyst we selected copper which can easily be interconverted between its two oxidation states, Cu^I and Cu^{II}. The normal geometry of Cu^I (tetrahedral) differs from that of Cu^{II} (planar). If a Cu^I–Cu^{II} system is to be used to catalyse fast electron transfer, the environment around both the two copper atoms must, therefore, be distorted. It is much easier to distort the tetrahedral geometry of Cu^I than to distort the planar Cu^{II} into a tetrahedron. By suitable choice of ligands, however, we can build various near-tetrahedral geometries around the Cu^{II} and then follow the rate of electron transfer. We have done this in the solid state—for example, by forcing copper(II) chloride into a lattice of copper(I) chloride with cobalt hexamine as a counter ion. The resulting dark crystals have electrical conductivity (i.e. electron transfer rates are relatively high). Another method of achieving a ground state which looks like the required transition state is to prepare organic molecules which will have strong steric hindrance to the formation of a planar environment around copper(II). The ligands are then caused to twist out of plane (this fact has been established by crystallography). In such compounds the rates of electron transfer are enhanced.

What has this got to do with enzymology? The crystal structure of the so-called blue Cu^{II} electron-transfer centres is now known from the (unpublished) work of H. Freeman (Sydney, Australia): Cu^{II}, instead of sitting in its normal planar geometry, has been distorted into a nearly-tetrahedral environment by two sulphur and two nitrogen ligands (for review see Williams

1971). Your approach is the equivalent in organic chemistry but for much harder reactions than these unimolecular electron transfers.

Knowles: It is a matter of great pleasure to me to see you stabilizing a transition state and not worrying too much about raising the level of ground states (cf. Vallee & Williams 1968)!

Williams: This is exactly what I do not do. It is exactly the entatic-state idea of adjusting a ground state that I am describing.

Knowles: But we are trying to stabilize transition states.

Williams: Yes, you are lowering the transition state of the substrate but you have improved the catalytic attack of your 'enzyme' by forcing a matching ground-state geometry in the protein model. One can either alter the geometry of the substrate on binding or improve the catalytic property of organic residues in the 'enzyme'. Your experimental method adjusts the energy of the required substrate geometry but it also forces like charges together in the catalyst—i.e. it plays chemistry in the same way as an enzyme, using both lowered transition states and raised ground states.

A different point is that I am not sure that your approach will give the symmetrical product because phosphate does not particularly like the symmetrical position you draw—it will move to one side of the hole, forming two hydrogen-bonds, and rattle round in the binding site. If you are not careful, the phosphate will be stabilized as a tetrahedron and it will ignore one third of the charges of the sugars. A direct analogy is with the lithium chloride lattice; the lithium ion does adopt a symmetrical position but wanders round the hole made by the chloride ions.

Knowles: The peripheral aminomethyl groups have a certain flexibility. They can be rotated so as to maximize the interaction (either charge–charge or hydrogen-bonding) with the oxygen atoms of the phosphoryl group. They can turn away or turn in. Our only real concern is that they lie in a plane. The flexibility of these aminomethyl groups will allow this interaction to be maximized. We then want to 'tune' the hydrophobic binding locus of the substrate to move the phosphorus up or down, into the plane of the amino groups. So I do not believe that the phosphate will rattle between three immobile amino groups, because they are not rigid; they are only restricted to a plane.

Richards: How specific is the 'fine tuning' with easily synthesizable compounds—0.5, 0.01 Å or what?

Knowles: It would be rash of me to give a precise figure. If, from molecular models and the available spectroscopic data, we should decide what is wrong with a host–guest complex that is, say, not decomposing rapidly, we could in principle do something about it.

Arigoni: In the synthesis of the symmetrically substituted cyclodextrin what was the yield of the symmetric compound relative to that of the other 18 possible isomers? Statistically speaking, your target compound is the least favoured.

Knowles: The isolated yield of the tritrityl compound is 15–20%. The total yield of tris-modified material in the mixture of tritrityl isomers is probably 70–80%. We chose the large trityl group hoping that vicinal detritylation would be disfavoured (i.e., reaction of the hydroxy groups adjacent to an already tritylated group would not occur easily). By careful selection of the reaction conditions we minimized the undesired products, after, of course, we had worked out which chromatographic spot was what.

Richards: In spite of everything that you have said why do you need the symmetrical compounds? Wouldn't a 'lather' of six charged groups serve your purpose just as well?

Knowles: But how inelegant! The hexamino compound could more readily be made and, protons being protons, one would imagine that they would rearrange themselves appropriately. But we wanted to be more explicit in the design of our host species.

This field of research, unfortunately, is full of 'gee-whiz' experiments—for instance, catalytic micelles can give rate enhancements of 10^6–10^8. But it seems to me that one is then replacing a problem, enzyme catalysis (which is well-defined), by one that is undefinable. This is also true for many polymeric molecules—e.g., polyethylenimine. In contrast, the aim of our exercise is to try to determine whether current ideas about enzyme catalysis can form a basis for the *rational* design of catalyst.

Blow: In studies of a locked substrate of chymotrypsin (which, admittedly, is a special case) we found that a 0.25 Å shift in the position of the ester group that was going to be hydrolysed made little difference to the rate constant, but a 0.5 Å shift caused changes in reactivity ranging from 15- to 2000-fold (Rodgers *et al.* 1976).

Knowles: It would be good if that turned out to be a general rule.

Vallee: The distances and measurements obtained with enzymes in terms of motion and sizes are probably not yet mapped adequately enough to enable us to design successful models. That is not to say that it will not be possible.

References

RODGERS, P. S., GOAMAN, L. C. G. & BLOW, D. M. (1976) On the correlation between three-dimensional structure and reactivity for a series of locked substrates of chymotrypsin. *J. Am. Chem. Soc. 98*, 6690-6695

VALLEE, B. L. & WILLIAMS, R. J. P. (1968) Metalloenzymes: the entatic nature of their active sites. *Proc. Natl. Acad. Sci. U.S.A. 59*, 498-505

WILLIAMS, R. J. P. (1971) *Inorg. Chim. Acta 5*, 137-155

Stereochemical studies of enzymic C-methylations

D. ARIGONI

Organisch-Chemisches Laboratorium, Eidgenössische Technische Hochschule, Zürich

Abstract Samples of methionine carrying a chiral CHDT-group have been prepared from and degraded chemically to chiral acetic acid. With this powerful tool
inversion mechanisms were detected for the methylations on sulphur, oxygen and
carbon atoms in the biosynthesis of methionine, loganin and cyanocobalamin,
respectively. Chirality of the methyl group of methionine is retained during
formation of the C-24 methyl group in ergosterol biosynthesis. A stereochemical
scheme for the unusual course of this reaction is presented and the current stage of
experiments aimed at its verification is discussed.

The involvement of S-adenosylmethionine as a biological methylating agent
in a variety of biosynthetic processes is well documented. Mechanistically, the
transfer reaction is best visualized as a substitution on the methyl group of the
alkylating agent by attack of an appropriate nucleophilic centre of the substrate undergoing alkylation, and many studies on the biosynthesis of
branched fatty acids and of sterols with a modified side-chain have indicated
that isolated $C=C$ double bonds have sufficient nucleophilic character for
participating in such alkylations (for reviews, see Lederer 1977; Goad *et al.*
1974).

The two main pathways for monoalkylation of the Δ^{24}-unsaturated ste-
roidal side-chain are illustrated in Scheme 1. The carbonium ion formed by
transfer of the methyl group to C-24 of the substrate can occasionally suffer
loss of a proton from one of the methyl groups adjacent to the electron-
deficient centre (path *a*), but more commonly it will evolve along path *b*
through a 1,2-hydride migration and subsequent proton loss to the 24-
methylene compound. In at least one organism both processes seem to
compete on a single enzyme surface (Goad *et al.* 1972). It is particularly
striking that, to date, no *bona fide* evidence has been obtained for the operation of path *c*, which would result in the generation of a Δ^{24}-unsaturated

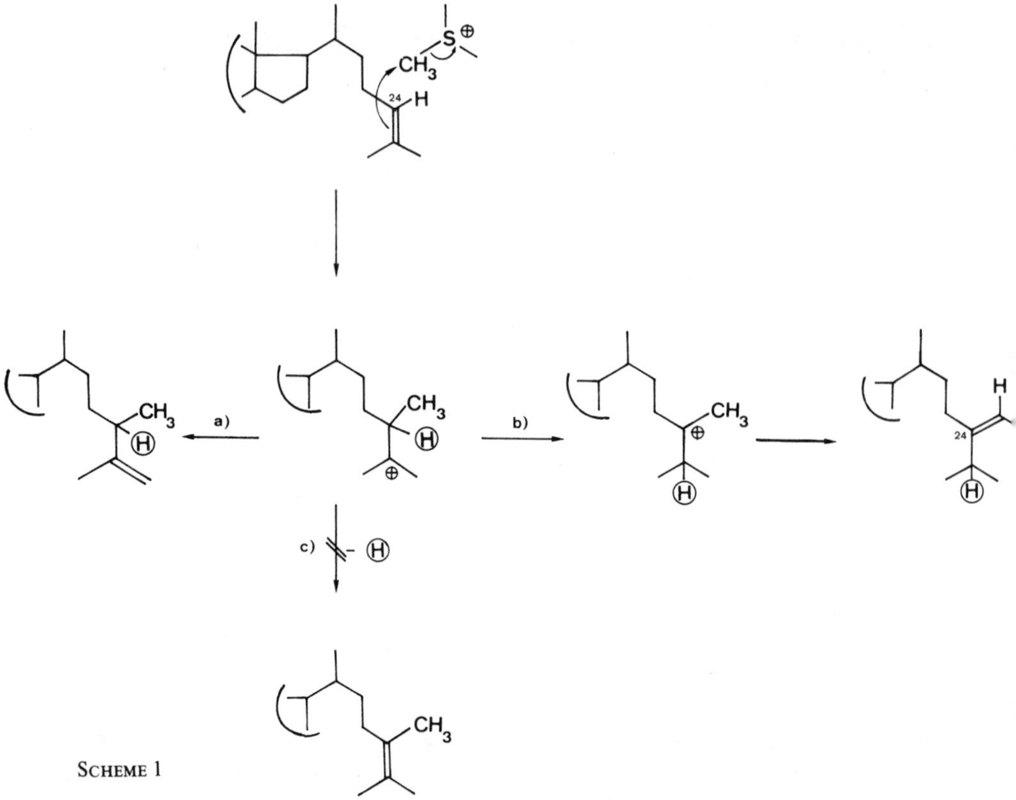

SCHEME 1

compound. Of the many speculative proposals which can be advanced to explain this peculiar state of affairs the one illustrated in Scheme 2 attracted our attention since it implies the possibility of uncovering the operation of a general principle. According to this simple hypothesis a negatively charged counter-ion located next to the sulphonium centre so as to facilitate its fixation at the active site of the enzyme is identical with the basic group ultimately responsible for abstraction of the proton. As a consequence, addition of the methyl group and removal of the proton must necessarily occur from the same face of the plane defined by the original bond of the substrate. With bridged structures as a visual aid for symbolizing the conformational stability of the ionic intermediates, it can be seen that formation of a Δ^{24}-unsaturated sterol from 2 is now inhibited by the fact that the bridging proton lies on the wrong side of the molecule, whereas deprotonation of 3 and 4 to the corresponding olefins is free from such a constraint. It may be mentioned in favour of the scheme that the postulated juxtaposition of methylating agent and proton-abstracting base is a prerequisite for the mechanistically related cyclopro-

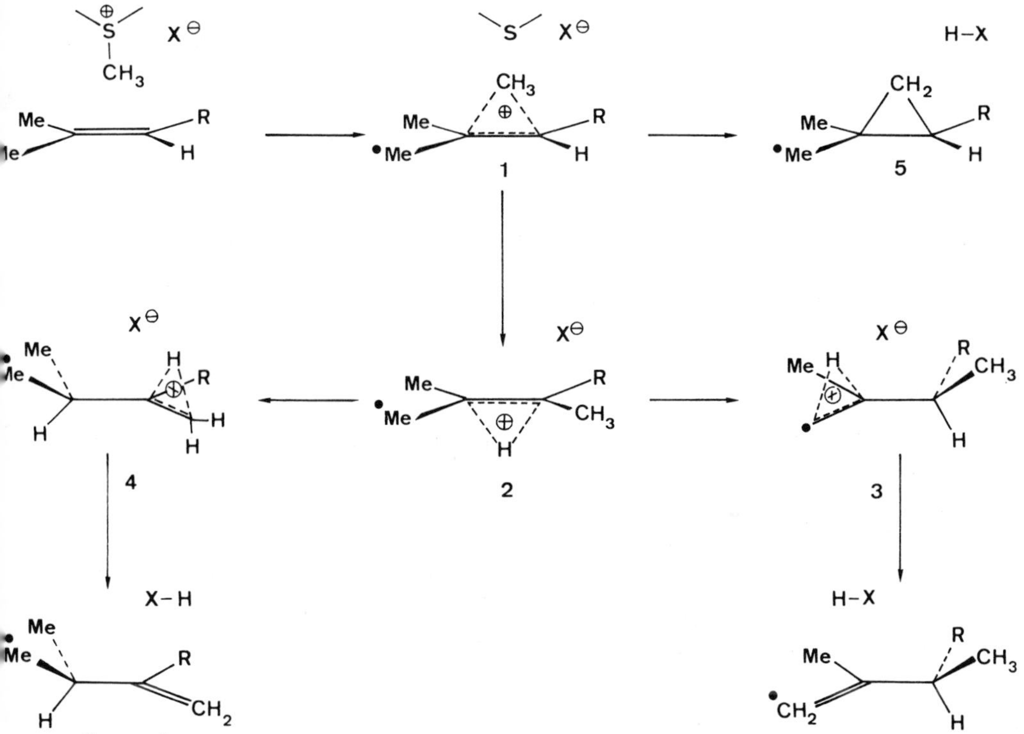

SCHEME 2

panation (cf. **1** → **5**) which has been detected in the biosynthesis of some fatty acids. However, the main merit of this hypothesis lies in the fact that an experimental approach can be devised to test its validity and the result of such experiments will at any rate cast a new and interesting light on the topological relationship between the substrate and the active site of the enzyme with which it interacts.

As I shall describe in detail further on, unravelling of the puzzle requires the availability of samples of methionine carrying a chiral CHDT-group of known configuration and I shall now digress from the main theme for a while in order to describe the preparation and some of the chemical properties as well as the biochemical behaviour of this crucial compound.

Chiral acetic acids of good enantiomeric purity and high tritium content are now available through a relatively simple operation (Townsend *et al.* 1975) and served as a suitable starting material. The acid was converted stepwise (Scheme 3) into the *p*-toluenesulphonimide of methylamine, which could be used as an alkylating agent for the facile conversion of homocysteine into methionine. A similar preparation of the same compound has been recently and independently reported by another group (Mascaro *et al.* 1977). In ap-

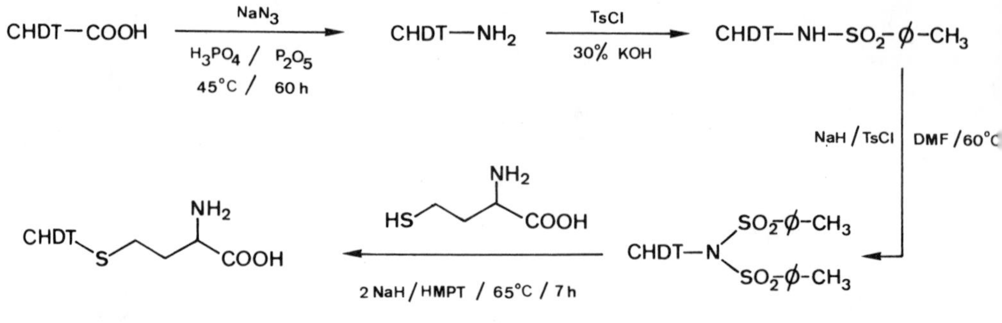

SCHEME 3

propriate control experiments it was demonstrated that no further transmethylation occurs between methionine and the anion of homocysteine, a
process which would have resulted in racemization of the critical methyl
group.

Characterization of the methionine samples so obtained proved more difficult than expected. After many abortive attempts at degradation, the
compound was converted into the methylsulphonium derivative (Scheme 4)

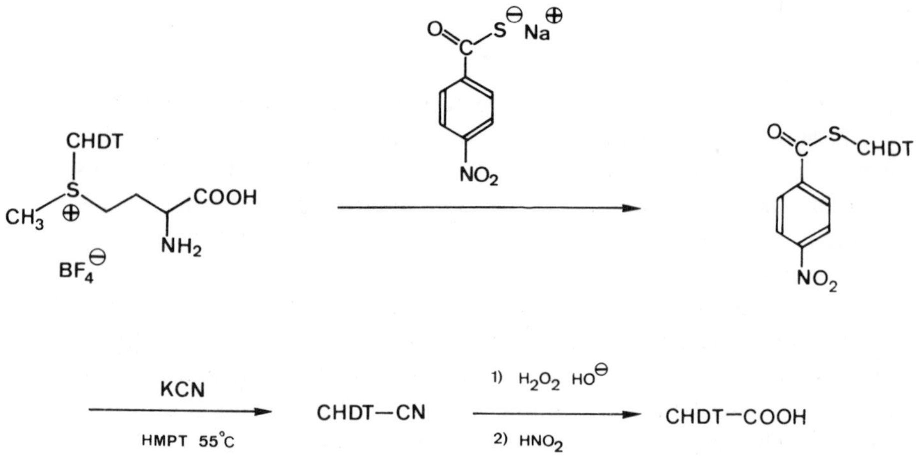

SCHEME 4

which could be easily demethylated with the anion from p-nitrothiobenzoic
acid. The resulting ester proved amenable to cyanolysis and the acetonitrile
was hydrolysed to acetic acid in carefully controlled conditions. The
enantiomeric purity of the different samples of doubly labelled acetate was

assayed in the usual way by converting enzymically their CoA-esters into malate and equilibrating the latter with fumarase (Cornforth *et al.* 1970; Lüthy *et al.* 1969). The F-values used for characterization of the samples refer to the percentage tritium retention detected in the equilibration step; expected values are F = 80 for the enantiomerically pure (*R*)-compound, F = 20 for the (*S*)-compound and F = 50 for the racemate. The complementary results obtained in two sets of experiments starting with antipodal materials (Scheme 5) indicate that the stereochemical integrity of the doubly labelled

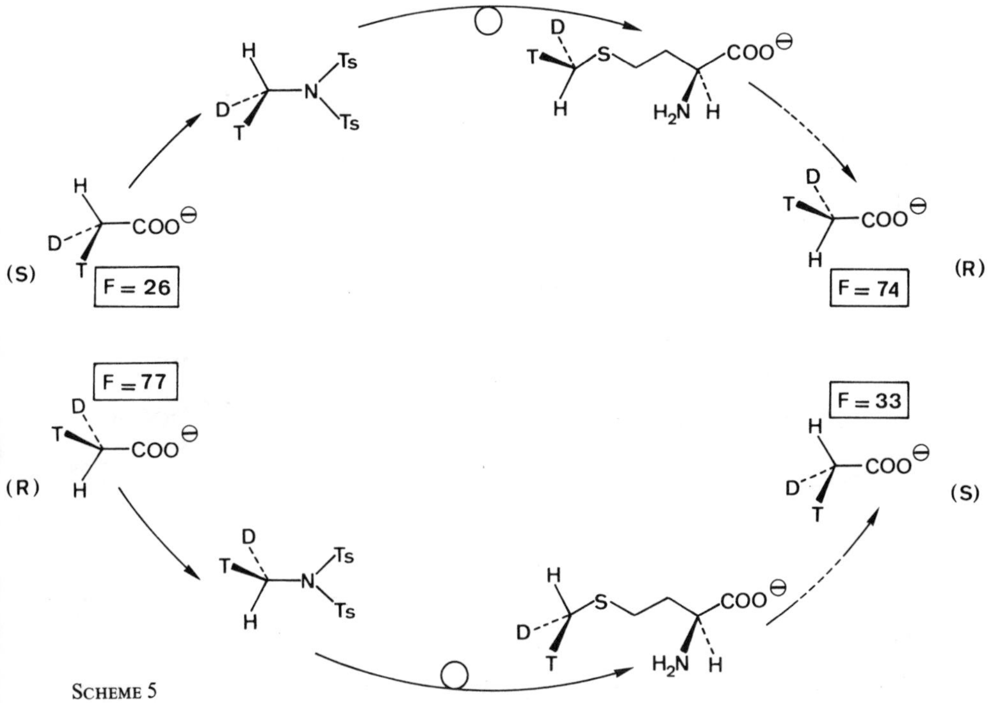

SCHEME 5

methyl group is maintained to a high degree in going from a given acetic acid to methionine and back to acetic acid. Thus, each of the steps of the overall conversion must be characterized by very high, if not complete stereospecificity. In addition, the overall inversion of configuration observed when going through one cycle is in excellent agreement with mechanistic expectations which predict three such inversions, one for each of the nucleophilic substitutions. Hence, it is safe to conclude that the methionine prepared from (*S*)-acetic acid carries a methyl group with (*R*)-configuration and *vice versa*.

With these very valuable tools in our hands we could now proceed to check
one of the central tenets in the theory of biological methylation, namely the
inversion of configuration at the centre which undergoes the alleged
nucleophilic substitution. At first we studied a case of sulphur-methylation
with an enzyme from Jack bean meal (Abrahamson & Shapiro 1965) which
catalyses a transmethylation from methylsulphonium-methionine to homocysteine (Scheme 6). The sole product of this reaction is methionine and it is

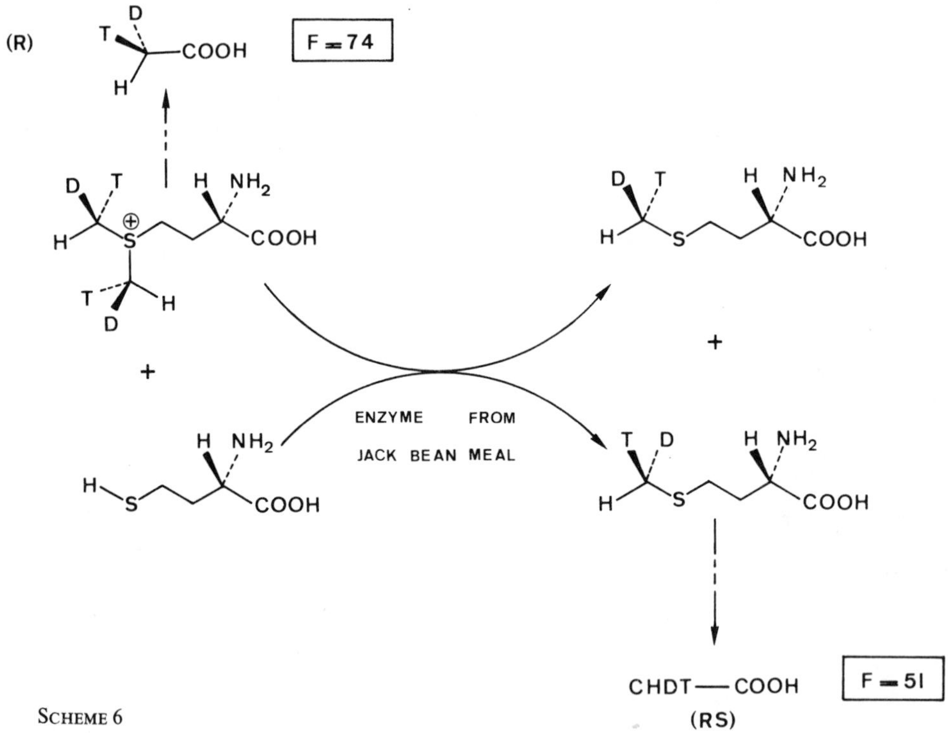

SCHEME 6   (RS)

important to note that half the molecules are derived from the demethylation
of the methylating agent, and the remainder stem from homocysteine. This
unique situation has a remarkable consequence. In the chemical preparation
of methylsulphonium-methionine from (R-methyl)methionine the alkylation
step is stereochemically random, so that the chiral methyl group of the
product will be equally distributed between the two possible diastereotopic
positions. Therefore, it does not matter whether the enzyme is able to differentiate between these two positions; at any rate, half the molecules in the
resulting methionine will retain the original configuration, whereas the configuration of the transported methyl group in the other half will depend

critically on the mechanism of the transfer reaction. In practice, degradation of the labelled methionine from a transmethylation experiment involving an alkylating reagent with (R)-methyl groups gave racemic, doubly labelled acetate, F = 51. Since the acetate from a similar chemical degradation of the alkylating agent retains its chirality, it must be concluded that, in accordance with expectation, the biological methyl transfer involves a net inversion of configuration. The connoisseurs will be pleased to note that this conclusion is independent of the knowledge of the configuration in the starting material!

Evidence for the same stereochemical course in a biological methylation on oxygen was obtained by following the formation of loganin from loganic acid on feeding of (R-methyl)methionine to the bitterweed, *Menyanthes trifoliata* (Scheme 7). The methyl group was recovered as acetic acid after cyanolysis of the labelled loganin and hydrolysis of the resulting nitrile. Again the results of the stereochemical assay are consistent only with an inversion mechanism for the alkylation step.

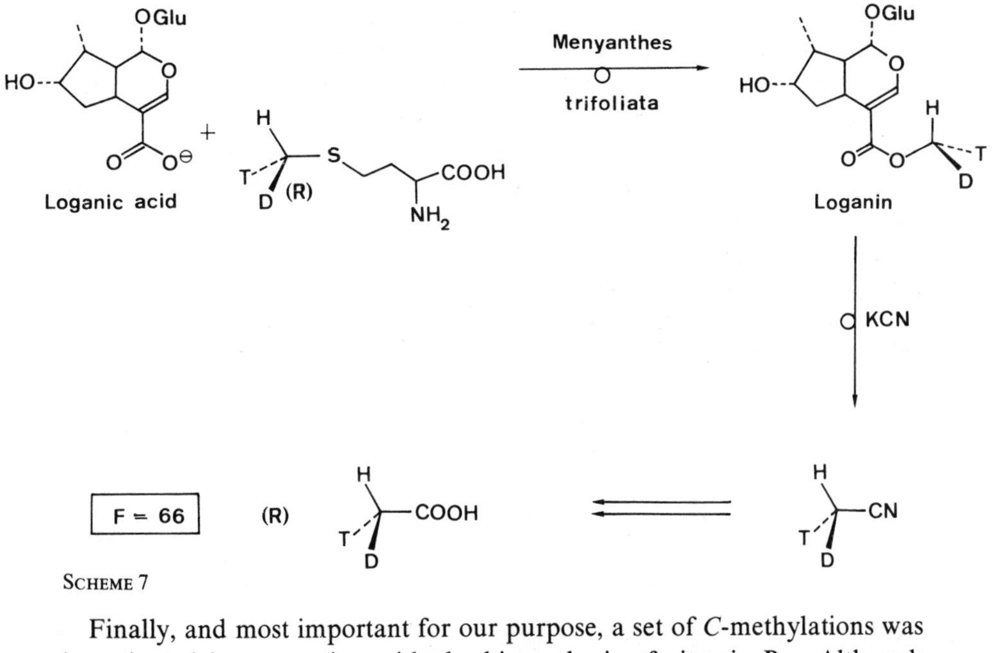

SCHEME 7

Finally, and most important for our purpose, a set of C-methylations was investigated in connection with the biosynthesis of vitamin B_{12}. Although several mechanistic details remain veiled, it is well established that formation of the fascinating array of atoms represented by cyanocobalamin (Scheme 8) proceeds from uro'gen-III in a complex sequence of steps which involves *inter alia* extrusion of C-20 and insertion of seven methyl groups (for a review see

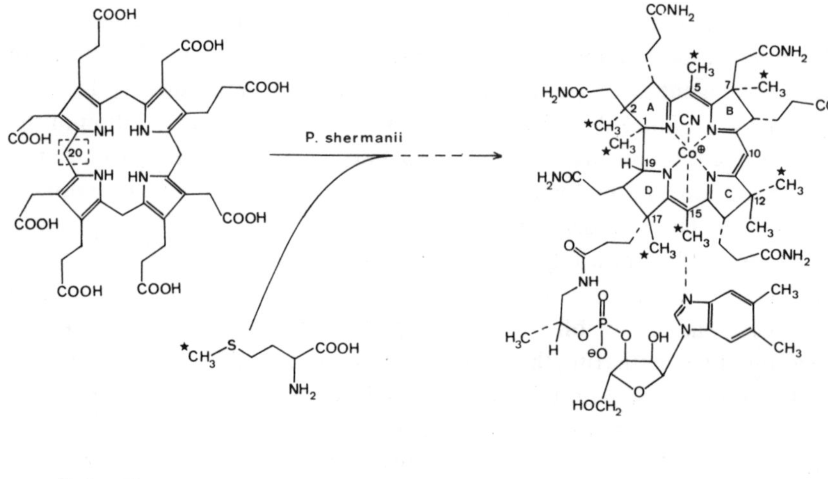

Uro'gen III   Cyanocobalamin
SCHEME 8

Scott 1975). These groups are known to originate from methionine and evidence has been provided independently from three laboratories that all seven methyl groups are transferred intact, i.e. without exchange of their protons (Imfeld *et al.* 1976; Battersby *et al.* 1976; Scott *et al.* 1976). Specimens of cyanocobalamin obtained from *P. shermanii* after the feeding of methionine carrying a chiral methyl group were converted into the cobester (Scheme 9) which was then degraded by ozonolysis to acetic acid (embodying the methyl groups at C-5 and, at least in part, at C-15) as well as to a set of three fragments corresponding to rings B and C and to the A–D-system of the starting material. Additional acetate samples were obtained from Kuhn–Roth oxidation of the B- and C-fragments. The results (condensed in Table 1) leave little doubt that in each case the *C*-alkylation had proceeded with inversion of configuration. A further example of inversion during biological *C*-alkylation has been detected recently (Mascaro *et al.* 1977).

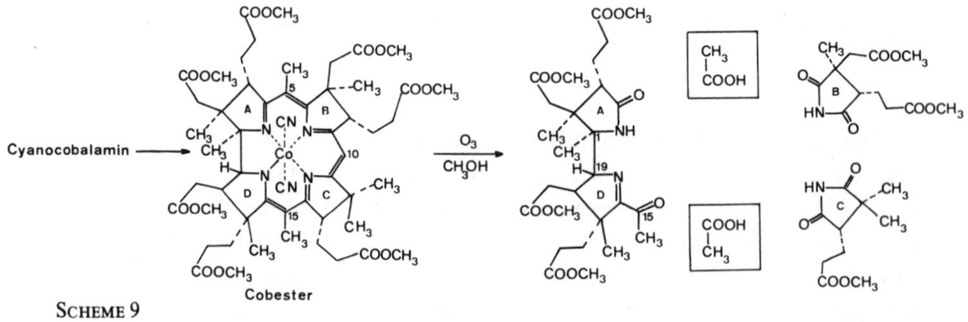

SCHEME 9

TABLE 1
F-values of acetate samples from degradation of cyanocobalamin

Precursor	(R-Methyl)-methionine	(S-Methyl)-methionine
Acetate from:		
C-5 (and partly C-15)	30	77
ring B	29	75
ring C	31	72

With the considerable support added by these experiments to the intuitive feeling that a linear transition state is a typical feature of biological methylations we can now revert to our main topic and take a closer look at the mechanism of side-chain alkylation in steroid biosynthesis. For practical reasons we elected to concentrate our efforts on the formation of ergosterol, a compound of widespread occurrence among lower organisms. The C-24-methyl group of ergosterol is generated by the hydrogenation of a 24-methylene intermediate, itself derived along path b of Scheme 1. At first glance, the inclusion of an additional hydrogenation step may seem to introduce an unnecessary complication into the main problem. In fact, this feature had purposely been taken into account as a specific bonus in the planning of the experiment. In a critical test of the possibility of putting the desired approach into practice methionine samples with a chiral methyl group were administered to actively-growing yeast, and it was rewarding to find that Kuhn–Roth degradation of the labelled ergosterols formed in these experiments gave chiral acetates which had retained the configuration while displaying an apparent enantiomeric purity corresponding to about 50% of the value of the starting material (Scheme 10). Two important conclusions can be drawn from this finding: (a) both the proton elimination which generates the

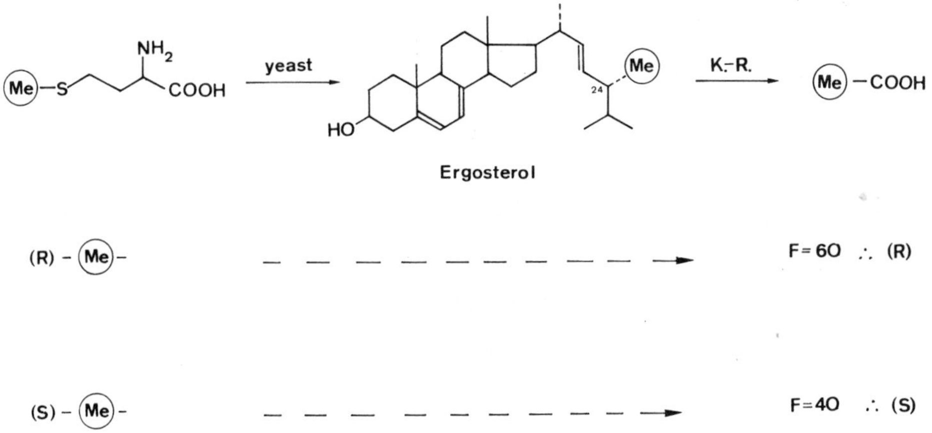

Ergosterol

(R) – (Me) –   _ _ _ _ _ _ _ _ _ →   F = 60 ∴ (R)

(S) – (Me) –   _ _ _ _ _ _ _ _ _ →   F = 40 ∴ (S)

SCHEME 10

24-methylene intermediate and the proton addition which triggers its subsequent hydrogenation must be subject to strict stereochemical control; (b)
since an inversion mechanism can be accepted safely (by now!) for the
transfer reaction, the overall stereochemical outcome ($R \rightarrow R$) demands that
protonation of the terminal methylene group in the hydrogenation step engages the face of the double bond opposite the one from which proton
removal had occurred during its formation. Accordingly, this behaviour
serves to characterize the two sets of labelled C-24-methylene intermediates,
although it fails to reveal their actual structure.

An additional comment is required by the apparent marked drop in the
enantiomeric purity of the acetates from the degradation experiments. This
result is not entirely unexpected and an important cause for it can be discerned easily by inspection of Scheme 11. After the transfer of a given chiral
methyl group and the hydride shift, the proton abstraction step can involve, in
principle, any of the three hydrogen isotopes. This in turn leads to the formation of three different and interrelated isotopic species for the 24-
methylene intermediate. Stereospecific protonation of the CTD-component

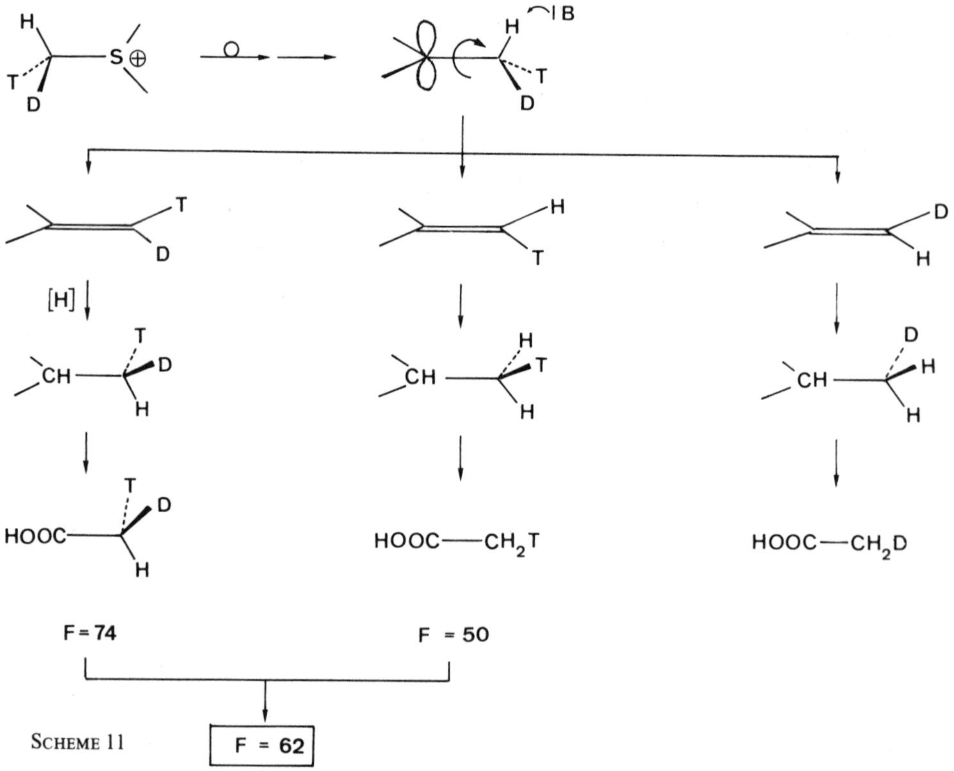

SCHEME 11

will eventually regenerate a chiral methyl group, whereas a similar protonation of the CHT-component will result in the formation of an achiral,
tritiated methyl group; on conversion into acetate, the latter will display in the
usual test an F-value of 50, thus simulating the behaviour of doubly labelled,
racemic acetic acid. In the absence of isotopic discrimination the F-values
expected for the presence of equal amounts of the two tritiated species fall
very short of the experimental results. It should be noted that these values are
not affected to a major extent by deuterium isotope effects which do not
exceed a value of 3–4.

Next we had to decide which of the two possible trajectories is followed
during the alkylation step proper. Since the subsequent hydride migration is
forced to take place on the opposite face of the original double bond (Scheme
12), this problem can be brought to a solution by labelling one of the methyl

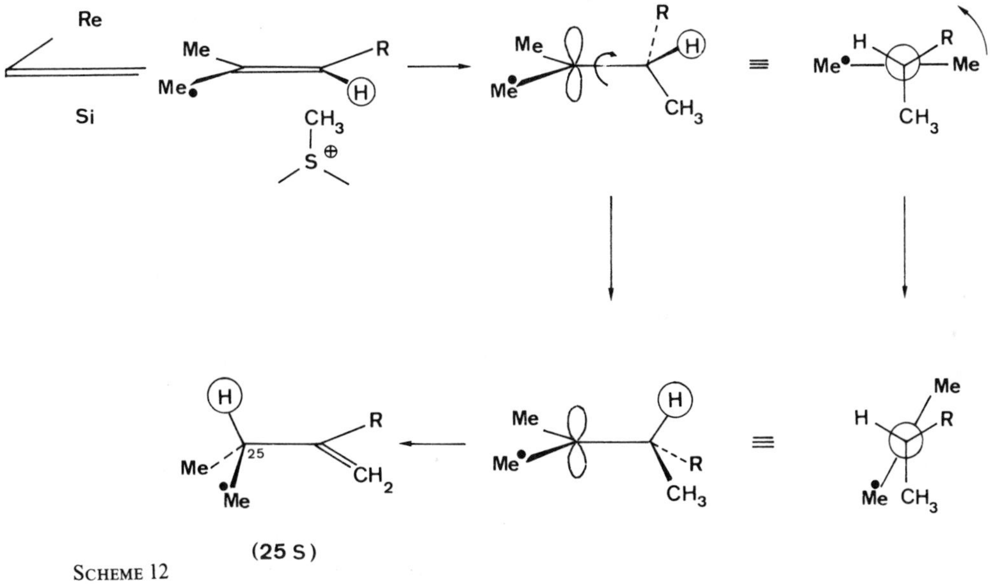

SCHEME 12

(25 S)

groups in the isopropylidene moiety of the precursor and determining the
chirality which develops at the terminus of the hydride migration. A sample of
lanosterol specifically labelled with ^{13}C in the E-methyl group seemed to serve
the purpose but, in spite of considerable experimentation, no appreciable
incorporation of this material into ergosterol could be observed on incubation
with yeast cells, probably as a consequence of adverse permeability barriers.
In the event this difficulty could be circumvented along a more indirect route
when it was found that a strain of *Claviceps paspali* is able, in one stroke, to

assemble the desired intermediate from [2-^{13}C]mevalolactone and to convert it into a mixture of ergosterol and its dihydroderivatives without undue dilution of the label (Scheme 13). The specific E-orientation of the critical methyl group in the labelled lanosterol follows from the known location of the label in the squalene precursor. The pure $\Delta^{8,14}$ monounsaturated derivative, obtained by catalytic hydrogenation of the ergosterol-containing fraction, displayed in its ^{13}C n.m.r. spectrum five well-enriched signals, of which only one belongs to a methyl carbon atom. This ensures that the identity of the E-methyl group of lanosterol is maintained on the way to ergosterol and hence that the hydride shift occurs stereospecifically.

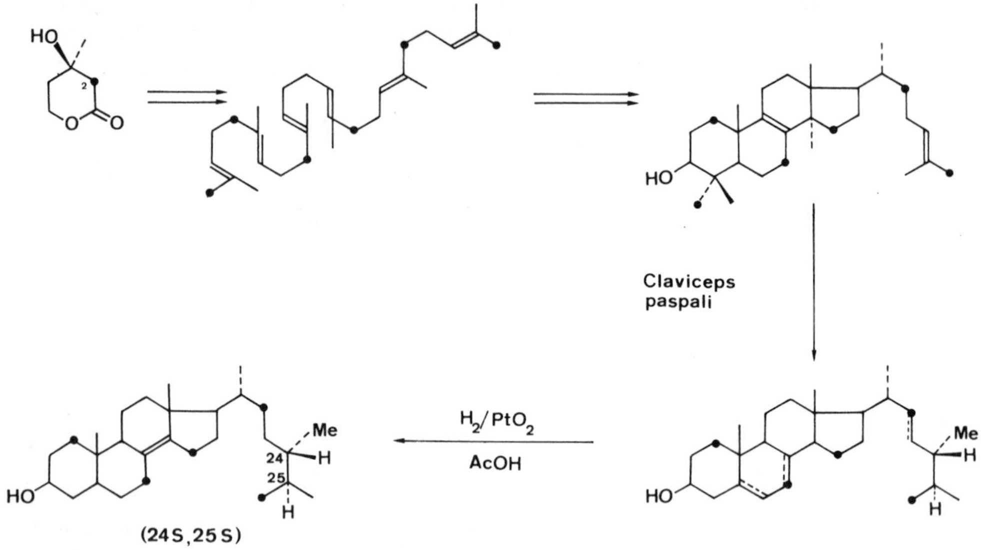

SCHEME 13

The C-25 configuration of the biosynthetic ergosterol derivative was determined by n.m.r. spectroscopic comparison with authentic material specifically labelled with ^{13}C in its Si-methyl group. The synthesis of this material is outlined in Scheme 14. The stereochemical outcome of the crucial Claisen rearrangement conforms with expectations (Ireland et al. 1976) and is verified independently by ozonolysis of the reaction product to a known compound. The perfect matching of the enriched methyl signal in the n.m.r. spectra of the synthetic and of the biosynthetic samples provides conclusive proof for the (25S)-configuration of the latter. From this and with reference to Scheme 12 it can be concluded that the methyl transfer in ergosterol biosynthesis engages specifically the Si-face of the substrate double bond.

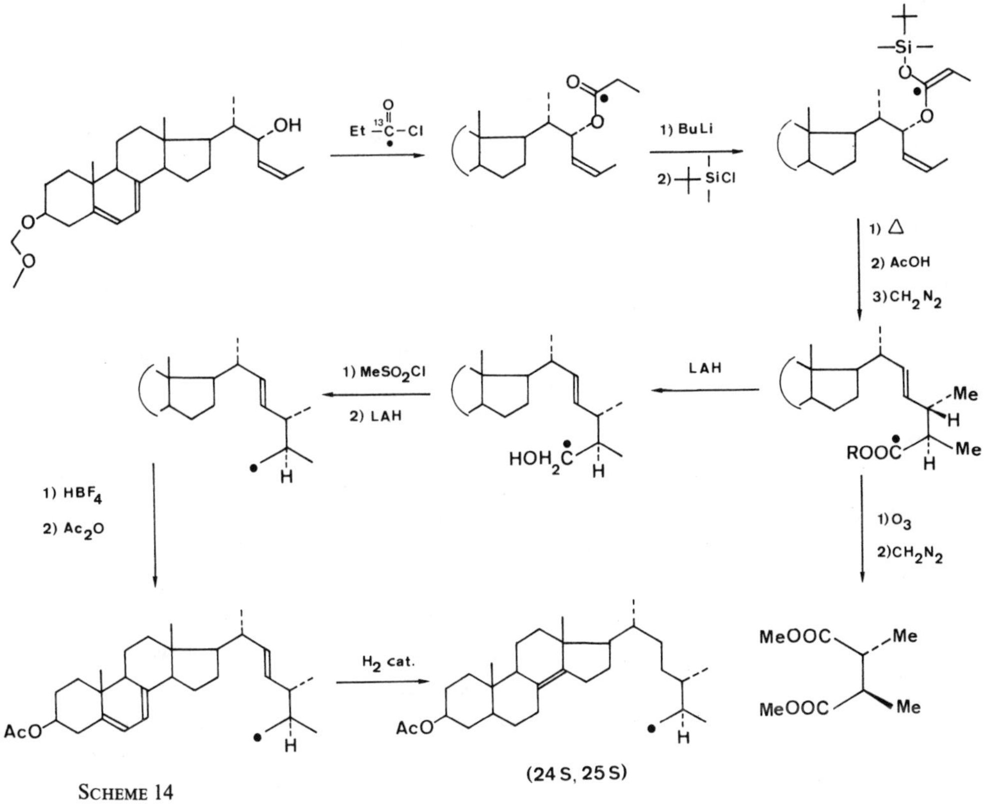

SCHEME 14

(24 S, 25 S)

To complete the story one needs only to establish the preferred stereochemical course for proton departure during formation of the elusive C-24-methylene intermediate. We had already seen by using a chiral methyl group that the process is indeed stereospecific and that the resulting doubly labelled intermediate can be characterized by looking at the chirality of the methyl group generated from it in the biological reduction. Thus the last bit of information can be arrived at by watching the specific formation of (R)- compared with (S)-chiral methyl groups on biological reduction of a synthetic, doubly labelled substrate with a known orientation of the two hydrogen isotopes in the terminal methylene group. Although this aspect of the study still awaits completion of the biological experiments, a satisfactory solution has been found for the non-trivial synthetic part of the problem (Scheme 15). By alternative use of a deuterium label in the reagent or in the substrate and n.m.r. spectroscopic analysis of the products it was a simple matter to show that the overall addition of chloromagnesium di-isopropyl cuprate (Westmijze *et al.* 1976) to the preassembled acetylenic compound of

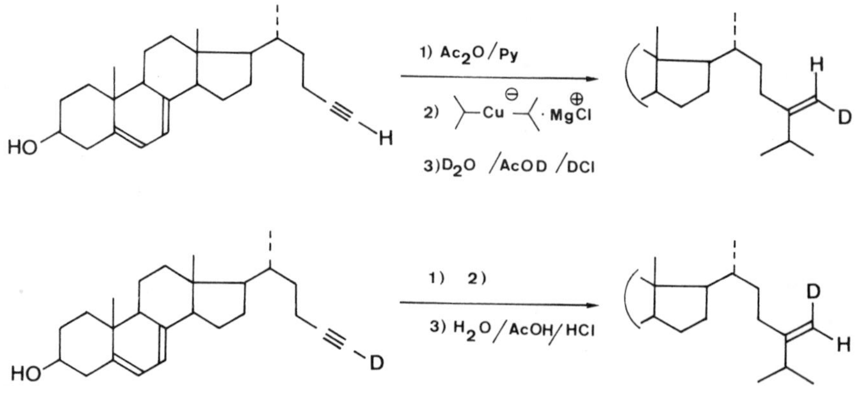

SCHEME 15

Scheme 15 occurs regio- and stereo-specifically in a *cis*-fashion. This paves the way for the unambiguous preparation of the corresponding doubly labelled stereoisomers.

The lack of information about the behaviour of these compounds as intermediates in ergosterol biosynthesis leaves us for the time being with an ambiguity as far as the original puzzle is concerned. The available evidence can be summed up as follows: (1) in each case investigated so far the biological transfer of methyl groups has been shown to proceed by an inversion mechanism; (2) in ergosterol biosynthesis, the methylation engages the *Si*-face of the substrate double bond; (3) in going from methionine to ergosterol through a methylene intermediate, a net retention of configuration is observed for the methyl group. These facts together with the well established configuration of ergosterol at C-24 impose severe restrictions on any stereochemical model and leave us with only two possibilities. If the hydrogenation step represents a *trans*-process, addition of the methyl group and proton loss are occurring on opposite faces of the original double bond. The overall transformation would be as depicted in Scheme 16 with a (*R*)-chiral methyl group in the methylating agent. Operation of this scheme would confront us with the problem of explaining how the migrating hydrogen manages to slide past the adjacent base without being captured by it! On the other hand a *cis*-mechanism of hydrogenation would imply that methyl donation and proton extrusion occur from the same *Si*-face of the double bond, in accordance with the original hypothesis. This would provide a welcome rationale for the necessity of the observed hydride migration. It should be recalled in this context that, for mechanistically related enzymic hydrogenations of non-activated double bonds, both *cis*- and *trans*-modes of addition

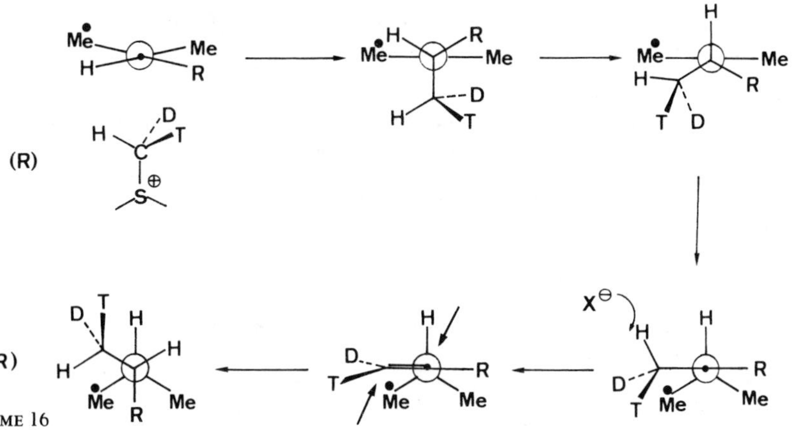

SCHEME 16

have been detected (for pertinent references see Watkinson *et al.* 1971; Rosenfeld & Tove 1971). Thus, the hydrogenation fails to conform to a unified stereochemical pattern and our problem must be left open on this account.

Although a final distinction between the two remaining possibilities will have to await the outcome of further experiments, it is felt that the main obstacles have been removed. The rest of the road looks clear and the solution within reasonable reach.

ACKNOWLEDGEMENTS

The work described in this presentation was made possible by the skillful, patient and dedicated collaboration of a group of younger colleagues. Dr Hugo Gorissen started the preparation of the labelled methionine and the project was completed by Philipp Huguenin who also investigated the methylation on sulphur and oxygen. Dr Craig Townsend and Marquard Imfeld took care of the experiment on vitamin B$_{12}$, and all the studies on ergosterol were done by Drs Philip DeShong and Pierre Le Roy. Last but not least, Dr Bruno Martinoni provided invaluable help in all the experiments with enzymes.

References

ABRAHAMSON, L. & SHAPIRO, S. K. (1965) The biosynthesis of methionine. Partial purification and properties of homocysteine methyltransferase of Jack bean meal. *Arch. Biochem. Biophys. 109*, 376-382

BATTERSBY, A. R., HOLLENSTEIN, R., McDONALD, E. & WILLIAMS, D. C. (1976) Biosynthesis of vitamin B$_{12}$. Evidence from double-labelling studies (^{13}CD$_3$) for intact incorporation of seven methyl groups. *J. Chem. Soc. Chem. Commun.*, 543-544

CORNFORTH, J. W., REDMOND, J. W., EGGERER, H., BUCKEL, W. & GUTSCHOW, CH. (1970) Synthesis and configurational assay of asymmetric methyl groups. *Eur. J. Biochem. 14*, 1

GOAD, L. J., KNAPP, F. F., LENTON, J. R. & GOODWIN, T. W. (1972) Observation of the side-chain alkylation mechanism in a *Trebouxia* species. *Biochem. J. 129*, 219

GOAD, L. J., LENTON, J. R., KNAPP, F. F. & GOODWIN, T. W. (1974) Phytosterol side chain biosynthesis. *Lipids 9*, 582

IMFELD, M., TOWNSEND, C. A. & ARIGONI, D. (1976) Intact transfer of methyl groups in the biosynthesis of vitamin B_{12}. *J. Chem. Soc. Chem. Commun.*, 541-542

IRELAND, R. E., MUELLER, R. H. & WILLARD, A. K. (1976) The ester enolate Claisen rearrangement. Stereochemical control through stereoselective enolate formation. *J. Am. Chem. Soc. 98*, 2868

LEDERER, E. (1977) in *The Biochemistry of Adenosyl-methionine* (Salvatore, F., Borek, E., Zappia, V., Williams-Ashman, H. G. & Schlenk, F., eds.), pp. 89-126, Columbia University Press, New York

LÜTHY, J., RÉTEY, J. & ARIGONI, D. (1969) Preparation and detection of chiral methyl groups. *Nature (Lond.) 221*, 1213

MASCARO, L., JR., HÖRHAMMER, R., EISENSTEIN, S., SELLERS, L. K., MASCARO, K. & FLOSS, H.G. (1977) Synthesis of methionine carrying a chiral methyl group and its use in determining the steric course of the enzymatic C-methylation of indolepyruvate during indolmycin biosynthesis. *J. Am. Chem. Soc. 99*, 273

ROSENFELD, I. S. & TOVE, S. B. (1971) Biohydrogenation of unsaturated fatty acids. *J. Biol. Chem. 16*, 5025

SCOTT, A. I. (1975) Concerning the biosynthesis of vitamin B_{12}. *Tetrahedron 31*, 2639

SCOTT, A. I., KAJIWARA, M., TAKAHASHI, T., ARMITAGE, I. M., DEMON, P. & PETROCINE, D. (1976) The methylation process in corrin biosynthesis. Application of $^1H[^{13}C]$ nuclear magnetic resonance difference spectroscopy to a biochemical problem. *J. Chem. Soc. Chem. Commun.*, 544-546

TOWNSEND, C. A., SCHOLL, T. & ARIGONI, D. (1975) A new synthesis of chiral acetic acid. *J. Chem. Soc. Chem. Commun.*, 921

WATKINSON, I. A., WILTON, D. C., RAHIMTULA, A. D. & AKHTAR, M. M. (1971) The substrate activation in some pyridine nucleotide linked enzymic reactions. *Eur. J. Biochem. 23*, 1

WESTMIJZE, H., MEIJER, J., BOS, H. J. T. & VERMEER, P. (1976) Stereospecific addition of heterocuprates to 1-alkynes in tetrahydrofurane. *Rec. Trav. Chim. Pays Bas 95*, 299-303

Discussion

Cornforth: I can add a pendant to this excellent work: the elucidation of the absolute configuration at sulphur of S-adenosylmethionine (Cornforth *et al.* 1977; the work was started by Paul Talalay in my laboratory at Milstead). The compound (1) presents considerable difficulty for the preparation of a crystalline substance suitable for X-ray crystallography. We decided to treat (S)-methionine (2) with iodoacetic acid (as had been done by Gundlach *et al.* 1959) to obtain a mixture of diastereomeric sulphonium salts (3) in equal proportions (Scheme 1). We could just separate the diastereomers on an amino acid analyser. That was not a preparative method but, by crystallization of the pentaiodide salts, we obtained both diastereoisomers (A and B) pure. We converted these into trinitrobenzenesulphonate salts, one of which (B) was suitable for X-ray crystallography and proved to have the (R)-con-

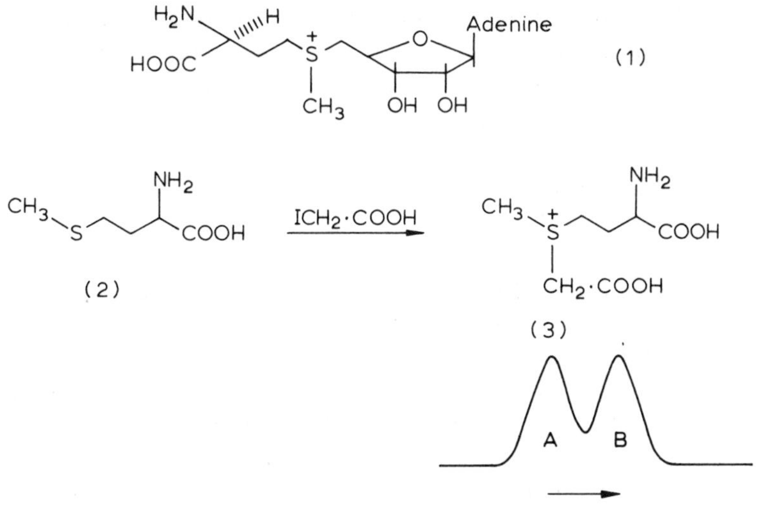

SCHEME 1 (Cornforth). Preparation of diastereomeric salts (3), showing result of ion-exchange chromatography of carboxymethylmethionine diastereomers.

figuration at the sulphur atom. Radioactive S-adenosylmethionine (10 μmol) was treated at pH 12 for 10 min at 25 °C. As expected, adenine was eliminated and after reducing the pH to about 4 and adding sodium periodate with titration to maintain constant pH we obtained, on chromatography, radioactive material which coincided with the elution time of A. The radioactivity from the radioactive fractions co-crystallized with the trinitrobenzenesulphonate of A but was almost completely lost from the trinitrobenzenesulphonate of B on crystallization. So we concluded that S-adenosylmethionine exists in the (S)-configuration at sulphur.

Battersby: Professor Arigoni, you have founds that ozonolysis of vitamin B_{12} clips out the A–D part of the molecule as one fragment and as Professor Eschenmoser had shown, the B and C rings are isolated separately. The A–D part has one methyl group still attached and you said that acetic acid was produced from both C-5 and C-15. What are the relative yields of acetic acid from these two methyl groups?

Arigoni: We cannot be certain as the total yield of acetic acid is only about 5% of the amount expected from two equivalents. Part of this acetic acid may be derived by further degradation of the A–D fragment.

Lipscomb: Does the exchange (of H⁺) with the surroundings always proceed through the bridged structure – the three-centre bond?

Arigoni: No. This is just one way of looking at it. I am not implying anything about the exact electronic nature of the intermediate, but it is a useful way to explain the maintenance of stereochemistry.

Lipscomb: Loss of a proton from a three-centre bridge leaves a perfectly good bond behind whereas if it is lost from a terminal position it leaves a lone pair. In the boron species the proton always comes from a bridge bond and not from the terminal bond. A carbonium ion, the compound you described, is analogous.

Knowles: How is the methyl group transferred to the double bond from *S*-adenosylmethionine?

Arigoni: Nobody really knows. If it could be shown that methyl donation and proton loss engage different faces of the substrate, as in Scheme 16, one might argue that at the active site of the enzyme the counter-ion has been removed from the sulphonium centre. The resulting insulation of the positively charged species could well provide additional driving force for the reaction. Consequently, even a weak nucleophile may interact with the methyl group.

Knowles: Are you happy with attack by *such* a weak nucleophile?

Arigoni: Can you suggest any other possibility bearing in mind that the reaction does occur? A chemical precedent has been reported (Chuit & Felkin 1967). The sulphonium compound (4) cyclizes cleanly to the bicyclooctanol (5) on heating in aqueous solution. This indicates that participation of double bonds in the solvolysis of sulphonium salts is feasible if the reaction centres are forced into close proximity. The biological methyl transfer reaction is nothing else but an intermolecular variation of this process and the enzyme may well play the trick simply by holding the two reaction partners within critical distance.

(4)   (5)

Williams: In B_{12}-containing enzymes the methyl group can be transferred onto cobalt, can't it?

Arigoni: Yes, very easily.

Battersby: Cobalt(I) is a powerful nucleophile.

Arigoni: We have attempted to prepare methyl cobalamin with a chiral methyl group and found that in the presence of the cobalt(I) species which served as a starting material the stereochemical integrity of the doubly-labelled methyl group is lost with a half life of less than one minute. The methyl group must be hopping around very fast indeed from one compound to the other.

Lipscomb: Bridging methyl groups are not unheard of.

References

CHUIT, C. & FELKIN, H. E. (1967) Alkylation intramoléculaire de la double liaison d'un sel de sulfonium insaturé. *C.R. Acad. Sci. Paris 264 (C)*, 1412-1413

CORNFORTH, J. W., REICHARD, S. A., TALALAY, P., CARRELL, H. L. & GLUSKER, J. P. (1977) Determination of absolute configuration of the sulfonium center of S-adenosylmethionine. Correlation with the absolute configuration of the diastereomeric S-carboxymethyl-(S)-methionine salts. *J. Am. Chem. Soc. 99*, 7292-7300

GUNDLACH, H. G., MOORE, S. & STEIN, W. H. (1959) The reaction of iodoacetate with methionine. *J. Biol. Chem. 234*, 1761-1764

General discussion

REACTIONS IN SUPERCOOLED WATER

Franks: For some time I have been fascinated by the possibility of uncoupling low temperature from freezing by making use of supercooled water. Fig. 1 shows the phase diagram for water. The extrapolation of the liquid/vapour curve is the supercooling curve. At the low-temperature end there is the curve for amorphous ice, which is obtained by the deposition of water vapour onto a very cold surface. There is a gap in the supercooling curve from the homogeneous nucleation temperature at −38 °C (at which water spontaneously nucleates itself and freezes) to the glass transition point for amorphous ice at about −138 °C. This gap is not at present accessible to experiment. Water can be supercooled only if heteronucleation (nucleation by impurities) is avoided. One way of reaching the homogeneous nucleation

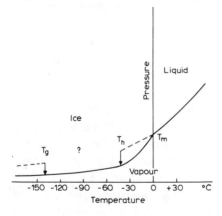

FIG. 1 (Franks). Phase diagram of water: T_m equilibrium melting point; T_h homogeneous nucleation temperature; T_g, glass transition of vitreous ice. The gap between T_g and T_h cannot be studied experimentally, because of the ultra-high cooling rates (about 10^6 degree/s) which are required for the vitrification of liquid water.

263

temperature is to emulsify water in an oil. We have put this approach to some use in the study of enzyme reactions. We have investigated three systems so far: cytochrome P450 and its substrate (Douzou *et al.* 1978), horseradish peroxidase (the recombination of the enzyme with carbon monoxide after photodissociation of the complex); and luciferase (the bioluminescent oxidation of flavin mononucleotide (P. Douzou *et al.*, unpublished results). We had three main objectives: first, to establish whether in supercooled water the systems showed the same kinetics as in mixed aqueous–organic solvents, secondly, to stabilize enzyme–substrate complexes and, thirdly, to try to accumulate labile intermediates. In fact, we have achieved each of these objectives. For instance, we succeeded in stabilizing the cytochrome P450–camphor complex, with the enzyme in the Fe(II), Fe(III) and Fe(II)CO states. The kinetics in supercooled water are not the same as in the mixed solvents. For example, the rate constant for the combination of horseradish peroxidase with carbon monoxide at $-30\,^{\circ}$C is 2×10^2 l mol^{-1} s^{-1} in super-cooled water but 1×10^{20} l mol^{-1} s^{-1} in ethanediol/water (1/1 v/v). The energy of activation also differs; in supercooled water it is the same as in aqueous solution at temperatures above 0 $^{\circ}$C, showing the emulsion to behave as a true aqueous system—there are no surface or adsorption effects.

Finally, we have accumulated labile intermediates: with cytochrome P450, we were able to enrich the enzyme substrate complex with the Fe(II).O_2 state which, at $-20\,^{\circ}$C, seems to be extremely stable. By diffusing oxygen into this emulsion the concentration of such an intermediate can be enhanced. The same is true for the reduced flavin mononucleotide in its oxygenated state: FMNH$_2$.O$_2$ is accumulated by the diffusion of gaseous oxygen into the emulsion at sub-zero temperatures.

These results demonstrate that one can now study enzyme–substrate, rather than enzyme–inhibitor complexes, even without using stopped-flow methods. One can, more or less, select the desired reaction rate and turn a fast reaction into a slow one. However, in some instances, $-38\,^{\circ}$C is still not sufficiently low a temperature. The gap between $-38\,^{\circ}$C and $-138\,^{\circ}$C can be bridged: the system then exists in the solid supercooled (vitreous) state. This is achieved by 'confusing' water to prevent it from freezing for as long as it takes to cross the temperature gap, for instance, by the use of antinucleating polymers. These polymers tend to 'organize' water molecules in ways which are incompatible with the organization in the ice structure. Various carbohydrates are good 'anti-nucleators', as are also various water-soluble synthetic polymers. In this way, we have prepared vitreous aqueous states (Franks *et al.* 1977). This is of importance for biological preservation, because freezing can be avoided—the *in vivo* cellular state can be 'petrified' (Skaer *et al.* 1977).

Lipscomb: Several groups have studied low temperature effects, including Professor Phillips' and Dr Steitze's in our laboratory on carboxypeptidase A crystals.

Phillips: Haas & Rossman (1970) used concentrated sugar solution to prevent freezing; that sounds like the carbohydrate polymer Professor Franks referred to.

Franks: Sugars and other low-molecular-weight compounds cause great problems in *in vivo* systems through their osmotic activity which the polymers lack.

Lipscomb: There is a danger that, if a reaction can proceed by one of two mechanisms differing slightly in activation energy, lowering the temperature might cause the reaction to switch from one mechanism to the other.

Gutfreund: That same problem arises when one uses substrates that react faster or slower than the normal substrate. It is always dangerous to change the rate too much, as the reaction may then proceed by a different mechanism.

Franks: The rate of association of horseradish peroxidase with carbon monoxide decreases uniformly in aqueous emulsion from $+25\ ^{\circ}C$ to $-35\ ^{\circ}C$, but in ethanediol/water mixtures the rates below $0\ ^{\circ}C$ diverge from those in aqueous solution at ordinary temperatures. It is hardly surprising that the organic solvent should affect processes involving conformational changes in proteins (Franks & Eagland 1975).

STUDIES OF ACTIVE SITES OF METALLOENZYMES USING PERTURBED ANGULAR CORRELATION OF γ-RADIATION

Johansen: My comments relate to Professor Vallee's points on the local conformations of carboxypeptidase A in solution (pp. 197-223) and also to some of the general questions which have been raised at this meeting about flexibility of protein structures and the discrepancies between X-ray crystallographic and spectroscopic studies. Previously no direct method was available by which we could measure the same property of an active site in an enzyme both in solution and in the crystalline state. However, the spectroscopic method of perturbed angular correlation (PAC) of γ-rays is essentially independent of the physical state of the material, and we have recently applied this method to studies of the active site of human carbonate dehydratase (Bauer *et al.* 1976, 1977). When two γ-rays are emitted from a radioactive nucleus (e.g. Cd^{111m}) in a cascade and detected with a coincidence spectrometer, the coincident counting rate depends strongly on the angle between their directions of propagation. This angular correlation can be perturbed by the interaction of the nucleus in its intermediate state with the electric field existing in the molecule. Thus the result of a PAC measurement is a time spectrum of the counting rate, exhibiting the development in time of the hyperfine interaction frequencies. Hence both structural (i.e. metal–ligand geometry) and dynamic information can be obtained. Recently we have used this method to investigate the active site properties of carboxypeptidase A both in the crystal form and in solution (C. Christensen, R. Bauer & J. T. Johansen, unpublished work, 1977).

Zinc-free apocarboxypeptidase A crystals were prepared by established procedures and the Cd^{111m}-enzymes were obtained by adding Cd^{111m} directly to a suspension of the apo-crystals. At pH 7.3 and 22 °C the experimental PAC spectrum of the crystals could be fitted theoretically by use of a quadrupole interaction frequency (ω) of 162 ± 2 MHz and a symmetry factor $\eta = 0.20 ±$ 0.06, where $\eta = (V_{xx} - V_{yy})/V_{zz}$ and V_{xx}, V_{yy} and V_{zz} are the field gradients at the nuclear site. Thus in the crystals there appears to be only one structure at the metal atom in the active site. In contrast, the PAC spectrum of the enzyme in solution demonstrated that at pH 7.3 and 22 °C at least two conformations are present, characterized by $\omega_1 = 160 ± 4$ MHz and $\eta_1 = 0.21 ± 0.16$ for one of the forms (probably similar to the form observed in the crystals) and a new form with $\omega_2 = 215 ± 17$ MHz and $\eta_2 = 0.39 ± 0.21$. So far we have collected data over the pH range 5.5–10 for crystals and solutions of Cd^{111}-carboxypeptidase and it appears that, from pH 5.5 to 9.5, only one form of the enzyme exists in the crystalline state. At higher pH values another species becomes apparent probably due to ionization of a water molecule on the cadmium atom. Between pH 5.5 and 7.5 we can detect at least four different species in solution dependent on pH. Furthermore at pH > 8 yet other species become apparent. Since these data are still only preliminary, we have not yet extended the calculations in detail, but in principle we can derive information about the geometry of the metal complex at the active site and we hope that the data will allow such calculations for each detected form of the enzyme. In conclusion, the PAC spectral data on Cd-carboxypeptidase A suggest that crystallization of the enzyme results in a static pH-independent conformation, whereas in solution, pH-dependent multiple conformations are present at the active site.

Williams: What order of perturbation of angle does this technique detect?

Johansen: About 5°. Further, we can detect changes in the environment of the cadmium atom up to about 0.4–0.5 nm from the metal.

Williams: So the conformers that you detect have stereochemistries in the active site changed by about 5°. Can you tell anything about the changes in bond length?

Johansen: Theoretically yes. At low pH we observe at least two forms but we don't know whether this is due to titration of one of the ligands or to changes in bond length as a result of titration of other groups in the active site. Only more detailed analysis of the data will tell us that.

Roberts: I recall that this γ-ray method was originally used for measuring rotational correlation times of proteins. How do you separate the rotational effects from the geometrical and electronic factors?

Johansen: In solution, if we did not increase the viscosity, we can only determine the rotational correlation time of the enzyme. Consequently, we used 50% sucrose solutions to decrease the tumbling of the molecule, and then both the rotational correlation time and the static quadrupole interactions can be determined.

Williams: So there are no orientation factors in the crystals?

Johansen: No. The sample consists of a randomly oriented suspension of crystals.

Phillips: Isn't conformation likely to be more similar in a salt-free crystal and in normal aqueous solution than in 50% sucrose?

Johansen: The presence of salt does not affect the PAC spectrum of the crystals. The only check we can make on the effect on the conformation of 50% sucrose is activity of the enzyme, which is not affected.

Phillips: But the enzyme is probably active in the crystal.

Lipscomb: To test that point, have you examined the crystal with pure water and with 50% sucrose and observed any difference?

Johansen: We have not done that.

Lipscomb: Does the γ-ray emitted from the nucleus give information about the square of the wave function at the nucleus, as in the Mössbauer effect?

Johansen: It corresponds to the quadrupole splitting in the Mössbauer effect.

Lipscomb: That means that this method responds to the electron density at the nucleus and, therefore, to just the *s*-orbital parts but not to orbitals which have a node at the nucleus. So although extremely sensitive to overall charge and ligands, the method is somewhat less sensitive to nearest neighbour geometry.

Williams: This method makes observations in less than 10^{-9} s but some rotational tumbling times are much longer, about 10^{-7} s, in solution. This large difference suggests that you will observe different orientations, not fast averaging. If the molecule is not isotropic and the crystals are powdered, the measurements could give information about the orientation of the molecules at the moment of measurement, not due, specifically, to differences inside a particular molecule.

Franks: You mean, the motions are not averaged.

Williams: Yes.

Lipscomb: If the method samples only the nearest neighbour up to 0.4–0.5 nm and if one takes a random sample of many oriented crystals, one will get the same result as if one averaged in solution.

Roberts: Not necessarily; in e.p.r. spectroscopy, for example, there is a clear distinction between the spectrum characteristic of an isotropically tumbling species and the 'powder pattern' one obtains from an amorphous solid. The superposition of spectra from all possible orientations is not the same as the spectrum averaged over all possible orientations.

Williams: One has to be careful about which overall ensemble average one considers.

Lipscomb: I should add that when we talk about crystalline state we should characterize it by a unit cell, because most crystals of carboxypeptidase A, for instance, sold commercially have different unit-cell dimensions from those on which we worked. At least three different crystalline forms are known. We did not work on the commercially available form.

Johansen: We have used the α-form which we prepared in the laboratory. We have not measured the unit cell parameters. However, both the α- and the γ-enzyme and their Val- and Ile-isoenzymes exhibited identical PAC spectra.

Vallee: Carbonate dehydratase has been investigated by this technique. Have the crystals as well as the solution been examined and are the results comparable?

Johansen: We have not studied the crystals of carbonate dehydratase, because we had difficulties in preparing crystals of this enzyme.

Lipscomb: As a result of Professor Vallee's discovery that Tyr-248 can bind to zinc, we looked at our map in more detail and found that a small percentage of the enzyme probably has that additional conformation. We feel that our enzyme crystals thus show two different conformations. Furthermore, raising the pH above about 8.0 destroys our crystals.

Johansen: Are the cross-linked crystals destroyed?

Vallee: I should like to make a point to resolve any confusion that might arise. I visualize practically all that has been discussed here in the terms used by Lindestrøm-Lang 25 years ago, namely that there is a distribution of various conformations of proteins, whose relative populations are governed by variable and unpredictable energy barriers. Also one could not say, in advance, how many of these states would exist. Discussion of this in terms of a difference in physical states of proteins is more an accident of observation than one of substance. However, our methods of observation have become so many and varied that more and more of this type of multiplicity of conformational difference is becoming apparent. Computer calculations sometimes bring clarification; Professor Richards' approach is a generalization, at least in terms of the shapes and forms at the active site, which might have some predictive capacity. Whether a particular protein may reasonably have one major conformation with several minor ones remains to be found out, but we have to look for it.

CONCLUDING REMARKS

Lipscomb: With respect to sorting out the different influences of environment on proteins, I suggest that the mechanisms that will probably be singled out (if they can be separated) are already known to organic chemists. I am reminded of A. N. Whitehead's comment that 'everything that is new has already been said by someone who did not discover it'. It seems to me that when we eventually find what these factors are, people will say 'we knew that all the time', but they don't know it now. This symposium has brought together scientists from widely different disciplines and they have communicated with each other exceedingly well. The protein chemists and model makers have discovered that, as the protein crystallographers knew all along from the lack of resolution and the high temperature factors, the structures of proteins are disordered in certain side groups and that the structures are not the precisely defined atomic systems that one needs if one is to build a model. Such structures need modification, as we have heard. The n.m.r. spectroscopists have 'come of age'. I was greatly pleased to hear about Professor Wüthrich's two-dimensional Fourier-Transform starting from the spin-echo experiments—that seems to be a remarkable advance. That and the use of lanthanide-shift reagents have gone very far towards addressing the question, what is the structure of proteins in solution as compared with that of proteins in the crystalline state? These two methods, X-ray diffraction and n.m.r are the only ones that provide large amounts of data but, on the other hand, they give no information about the kinetics or the mechanism. Consequently, we have witnessed the development of several spectroscopic techniques and kinetic techniques help us to elucidate the mechanism. Turning to models, we have heard less about them than I had hoped for, but within the limits of the symposium we cannot do everything. It is unfortunate that Professor T. C. Bruice (who was invited) was unable to attend; he has some excellent models which simulate the positions of proton donors, nucleophiles and other reactive groups in the active sites of enzymes. In this respect, however, I was delighted to hear of Professor Knowles' efforts: construction of the hydrophobic part of a molecule together with some of the functional parts is very important. Supposing it is true, as Professor Vallee said, that the models are always bound to be different, we want to know why models are different. Perhaps not all of them will be different from the way enzymes work (when we discover how they do work). The meeting was also noteworthy in that most people did not give standard conference papers but were looking to the future rather than the past. Perhaps, there is no better example than the way Professor Blundell held this own in the discussion after his paper.

I shall close by referring to those physicists who say that there is nothing new, in principle, in chemistry since the wave equation solves all chemistry! However, it is appropriate to recall that Feynman, a physicist himself, in his *Lectures on Physics* (1970), commented that this view is akin to the statement that, if one knows the rules of movement of the pieces in chess, one knows all the strategy of chess. We may know the general principles by which enzymes work but we do not know how proteins are activated or how environment can modify groups and create special functions—but we shall find out. This symposium underlines the substantial progress being made.

References

BAUER, R., LIMKILDE, P. & JOHANSEN, J. T. (1976) Low and high pH form of cadmium carbonic anhydrase determined by nuclear quadrupole interaction. *Biochemistry 15,* 334-342

BAUER, R., LIMKILDE, P. & JOHANSEN, J. T. (1977) Metal coordination geometry and mode of action of carbonic anhydrase. Effect of imidazole on the spectral properties of Co(II) and ^{111}Cd(II) human carbonic anhydrase B. *Carlsberg Res. Commun. 42,* 325-339

DOUZOU, P., DEBEY, P. & FRANKS, F. (1978) Supercooled water as medium for enzyme reactions at subzero temperatures. *Biochim. Biophys. Acta 523,* 1-8

FEYNMAN, R. P. (1970) in *Lectures in Physics* (Feynman, R. P., Leighton, R. B. & Sands, M., eds.), Addison-Wesley, Reading, Massachusetts

FRANKS, F. & EAGLAND, D. (1975) The role of solvent interactions in protein conformation. *Crit. Rev. Biochem. 3,* 165-217

FRANKS, F., ASQUITH, M. H., HAMMOND, C. C., SKAER, H. LEB & ECHLIN, P. (1977) Polymeric cryoprotectants in the preservation of biological ultrastructure I. *J. Microscopy 110,* 223-238

HAAS, D. J. & ROSSMAN, H. G. (1970) Crystallographic studies on lactate dehydrogenase at −75 °C. *Acta Cryst. B. 26,* 998-1004

SKAER, H. LEB, FRANKS, F. ASQUITH, M. H. & ECHLIN, P. (1977) Polymeric cryoprotectants in the preservation of biological ultrastructure. *J. Microscopy 110,* 257-270

Index to contributors

Bold *entries refer to papers; others refer to contributions to discussions*

Indexes compiled by William Hill.

Subject index